普通高等职业教育计算机系列教材

Java EE 软件开发案例教程

（Spring+Spring MVC+MyBatis）

熊君丽　主　编

曾珍珍　牛德雄　李鸿宾　刘　鑫　副主编

電子工業出版社

Publishing House of Electronics Industry

北京 • BEIJING

内 容 简 介

本书按照学生信息管理系统的整体构建和业务功能实现的工作化过程设计各个章节的内容。本书将知识点逐步抽丝剥茧，形成由简到繁的任务和案例，推动项目主体部分的最终实现。本书的特点是实用性强、操作思路明晰。书中的案例程序由业界颇为流行的 IntelliJ IDEA 平台编译，项目基于 Gradle 工具构建，在 Tomcat 服务器上运行。项目案例使用小型轻便的 MySQL 数据库，集成了 Bootstrap 前端框架，后台采用流行的 Spring MVC、MyBatis、Spring 三大框架集成。

全书分为 8 章：第 1～3 章为项目案例的开发平台搭建部分，着重介绍了 IDEA 平台、Gradle 工具、Bootstrap 前端框架的集成；第 4～5 章为项目的功能实现部分，着重介绍了 Spring MVC 和 MyBatis 的使用细节；第 6～7 章为 Spring 容器的原理性介绍，分析了三大框架的集成原理；第 8 章拓展了 Spring Boot 的开发全过程，还介绍了 Maven 构建工具的使用。本书实现了学生信息管理项目的主要功能，提供了丰富的教学案例和实现细节，并以附录的形式展示了数据库的详细设计。

本书可以作为高等职业院校计算机软件专业学生的教材，也可以作为 Java EE 开发的培训教材，还可以作为从事 Java EE 软件开发工作的技术人员的参考用书。

图书在版编目（CIP）数据

Java EE 软件开发案例教程：Spring+Spring MVC+MyBatis / 熊君丽主编. —北京：电子工业出版社，2020.6
普通高等职业教育计算机系列规划教材

ISBN 978-7-121-38996-2

Ⅰ. ①J… Ⅱ. ①熊… Ⅲ. ①JAVA 语言－程序设计－高等职业教育－教材 Ⅳ. ①TP312.8

中国版本图书馆 CIP 数据核字（2020）第 073491 号

责任编辑：徐建军　　　特约编辑：田学清
印　　刷：北京捷迅佳彩印刷有限公司
装　　订：北京捷迅佳彩印刷有限公司
出版发行：电子工业出版社
　　　　　北京市海淀区万寿路 173 信箱　　邮编：100036
开　　本：787×1092　1/16　印张：14.5　　字数：380.5 千字
版　　次：2020 年 6 月第 1 版
印　　次：2024 年 1 月第 8 次印刷
定　　价：43.00 元

凡所购买电子工业出版社图书有缺损问题，请向购买书店调换。若书店售缺，请与本社发行部联系，联系及邮购电话：（010）88254888，88258888。

质量投诉请发邮件至 zlts@phei.com.cn，盗版侵权举报请发邮件至 dbqq@phei.com.cn。

本书咨询联系方式：（010）88254570，xujj@phei.com.cn。

前 言

关于本书

在很多开源项目中，Spring 框架的“全家桶”使用形式已广为流行。Java EE 框架种类繁多，十年间优胜劣汰，但 Spring 框架却以扎实的基础逐步开花结果，枝繁叶茂。MyBatis 也是一款优秀的持久层框架，几乎省去了所有 JDBC 代码和参数的手工设置及结果集的检索等操作。Spring 和 MyBatis 的完美结合构成了快速构建用户应用系统的软件框架集合。

在教学过程中，编者深深体会到学生对框架实战的整体面貌和开发细节难以把握，然后萌生了解构实战项目，再重构成教学知识点的想法，于是就有了本书。本书只有 8 章，200 多页，却涵盖了基础开发的全过程。

本书面向的读者

本书可以作为高等职业院校计算机软件专业学生的教材，也可以作为 Java EE 开发的培训教材，还可以作为从事 Java EE 软件开发工作的技术人员的参考用书。在学习本书之前，读者最好已经掌握了 Java 语言编程、JSP 和数据库的基础知识。

本书的结构

第 1 章通过一个案例展现了在 IntelliJ IDEA 平台开发 Web 应用的整个过程。通过这个案例，读者可以很快进入框架应用开发的世界。

第 2 章讲解了 Gradle 构建工具的安装配置，以及如何在 IntelliJ IDEA 平台下创建一个基于 Gradle 的项目并管理和部署项目运行。

第 3 章讲解了如何将前端框架 Bootstrap 和 EasyUI 集成到第 2 章用 Gradle 构建的项目中。

第 4 章针对 Spring MVC 框架设计了丰富的案例，详细介绍了与第 3 章项目的集成，以及请求映射和参数传递、校验和异常处理等实用性知识。

第 5 章讲解了 MyBatis 框架与前几章的总集成，通过大量示例介绍 MyBaits 框架的运用，实现了对数据库的各种简单操作和多表关联等复杂操作。

第 6 章讲解了 Spring 框架的核心 IoC 的原理及其在项目中的运用，也对 IoC 的各种配置方式进行了深入的剖析。

第 7 章讲解了 Spring 框架的 AOP 知识，对如何配置 AOP 给出了详细的步骤，还介绍了基于 XML 和基于注解的事务管理配置。

第 8 章是拓展知识，引入了业界较为流行的基于 Maven 构建工具的 Spring Boot 开发模式，用一个完整的案例展示了从搭建框架到实现前端对数据库访问的全过程。

如何使用本书

首先，安装虚拟机 JDK、Web 服务器 Tomcat、数据库 MySQL、数据库前端工具 Navicat。然后，安装 Gradle 软件作为项目的构建工具。最后，安装 IntelliJ IDEA，这也是最重要的。

如何下载代码&如何与作者联系

为了方便教师教学，本书配有电子教学课件及相关资源，读者可以使用如下网址下载本书的代码：

- 华信教育资源网（www.hxedu.com.cn），免费注册后下载。
- 码云：https://gitee.com/roadflower/SSM.git。

如有问题，可以在网站留言板留言或与电子工业出版社联系（E-mail: hxedu@phei.com.cn）。也可以通过电子邮箱 2556214@qq.com 与编者联系，或者进入 QQ 交流群（831269404）获取更多学习资源。

感谢

本书由广东科学技术职业学院的熊君丽担任主编，并编写了第 4～8 章；牛德雄编写了第 1 章和附录 A；曾珍珍编写了第 2 章；李鸿宾编写了第 3 章，并参与了本书的构思和校对；刘鑫负责项目代码的编写。要特别感谢牛德雄老师，他的引领给予了我提笔写书的勇气。学校合作企业珠海爱浦京软件股份有限公司实训邦团队也给予了技术支持，在这里一并表示感谢。

由于时间仓促，框架开发知识点过于丰富，书中难免存在疏漏和不足之处，恳请广大专家和读者给予批评和指正。

编　者

目 录

第1章

Java EE Web 项目开发平台 IntelliJ IDEA

本章学习目标

- 了解 SSM 框架基础知识
- 熟悉 IntelliJ IDEA 平台
- 掌握在 IntelliJ IDEA 平台下创建 Web 项目的方法

本章简单介绍 Java EE 软件开发的相关概念，以及在 Java EE 体系中如何进行基于 SSM 框架的软件开发。本章还介绍了 IntelliJ IDEA 平台的搭建；并通过一个 Web 项目案例的创建与运行，介绍了如何在 IntelliJ IDEA 平台下进行 Java EE Web 项目开发，为后面完成学生管理系统做准备。

1.1 Java EE 开发及 SSM 框架简介

1.1.1 Java EE 简介

Java EE（Java Enterprise Edition，Java 企业版）是在 Java SE（Java Platform Standard Edition，Java 开发工具标准版）（Java 的核心）基础上构建而成的，它有自己的架构体系，多用于企业级开发。Java EE 主要包括 Java Web 应用开发技术组件。这些技术包括表示层技术、中间层技术、数据层技术。Java EE 还涉及系统集成的一些技术。所有这些技术组件构成了 Java EE 架构体系。

本书采用 IntelliJ IDEA 平台进行 Java EE Web 项目的开发，重点介绍基于 SSM 框架的 Java EE 项目开发技术。

1.1.2 MVC 设计模式与 SSM 框架

1. MVC 设计模式简介

MVC 是一种设计模式。设计模式是经实践反复检验的设计经总结提炼而形成的一种模式。而 MVC 设计模式是将软件的代码分为 M、V、C 三层来实现的一种设计方案，是上述三层程序结构的一种具体实现。

MVC是Model-View-Controller的缩写，即模型（Model）-视图（View）-控制器（Controller）。它将业务逻辑、数据与界面显示分离，将业务逻辑处理放到一个部件里面，而将界面及用户围绕数据展开的操作单独分离出来。

MVC 强制性地使模块中的输入、处理和输出分开，使它们各自处理自己的任务。MVC减弱了业务逻辑接口和数据接口之间的耦合，并使视图层更富于变化。

1）模型

模型（Model）表示业务数据和业务规则。在 MVC 的三个部件中，模型拥有最多的处理任务。例如，它可以封装数据库连接组件、业务数据库处理组件，这样的模型能为多个视图提供数据。由于应用于模型的代码只需写一次就可以被多个视图重用，所以能提高代码的重用性。模型一般用 JavaBean 技术实现。

JavaBean 是一种由 Java 语言写成的可重用组件。JavaBean 类必须按照一定的规范编写，它通过提供符合一致性设计模式的公共方法设置或获取成员属性。换句话说，JavaBean 就是一个 Java 类，只不过这个类要按一些规则来编写，如类必须是公共的，有无参构造器，要求属性是 private 且需通过 setter/getter 方法取值等。按这些规则编写了之后，这个 Java 类就是一个 JavaBean，它可以在程序里被方便地复用，从而提高开发效率。

MVC 的模型层，就是由这些 JavaBean 构成的模型组成的，它们在服务器端承担了大部分复杂的计算工作。只有在控制器控制下其结果才能使用并在视图中展现。

2）视图

视图（View）是用户看到并与之交互的界面。对老的 Web 应用程序来说，视图就是由 HTML 元素组成的界面。在新的 Web 应用程序中，HTML 依旧在视图中扮演着重要的角色，但也有一些层出不穷的新的技术，如 Adobe Flash 和 XHTML、XML、WML 等一些标记语言。JSP 作为动态网页常常充当 Web 应用的视图。

MVC 的优点是能为应用程序处理很多不同的视图。在视图中其实没有真正的处理发生，它只与客户端交互，包括获取用户请求、向控制器传递数据、接收控制器传回的封装数据，以及展示数据。

3）控制器

控制器（Controller）的功能是接收用户的输入并调用模型和视图去满足用户的需求，所以当单击网页中的超链接和发送 HTML 表单时，控制器不需做任何处理。它只是接收请求并决定调用哪个模型构件去处理请求，然后确定用哪个视图来显示返回的数据。

典型的 MVC 设计模式有基于 Struts、Spring、Hibernate 框架（简称 SSH 框架）与基于 Spring、Spring MVC、MyBatis 框架（简称 SSM 框架）等模式。它们是一种基于 MVC 模式不同层的实现技术，即通过 SSM 等框架可以方便地实现 MVC 模式应用程序的不同层。

2．SSM 框架简介

框架（Framework）就是已经开发好的一组软件组件，程序员可以利用它来快速地搭建自己的应用软件。软件框架就好像建筑物的骨架和大的构件，由它们来构建应用软件系统，要比通过一条一条语句进行编程更快速。

从技术角度看，框架是整个或部分系统的可重用设计，表现为一组抽象构件及构件实例间交互的方法。

利用框架技术不仅能快速地构建自己的软件项目，而且能克服自编软件的某些缺陷，因为在设计软件架构时，设计者已经对其进行了充分的考虑（如负载均衡等）。所以，目前采用框架技术进行软件开发是程序员的首选。

SSM 框架（SSM 框架集）就是 Spring、Spring MVC、MyBatis 三个开源框架组成的框架集的简称，它们是在 Java EE 基础上创建的用于快速构建用户应用系统的软件框架集，在 Java 企业级应用系统开发中得到了十分广泛的使用。

其中：

- Spring 是一个轻量级的控制反转（IoC）和面向切面编程（AOP）的容器框架。
- Spring MVC 分离了控制器、模型对象、分派器及处理程序对象的角色，这种分离让它们更容易被定制。
- MyBatis 是一个支持普通 SQL 查询、存储过程和高级映射的优秀持久层框架。

SSM 框架与 Java EE 及应用程序的关系如图 1-1 所示。

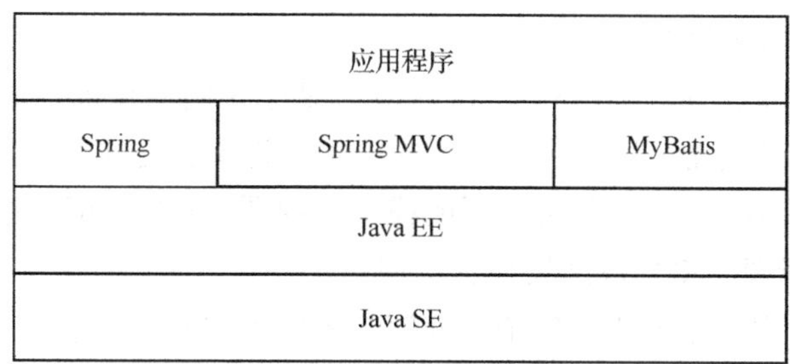

图 1-1　SSM 框架与 Java EE 及应用程序的关系

图 1-1 显示了 Java EE 是 SSM 框架的基础，而 Java SE 是 Java EE 的基础。前面已述，Java EE 提供了大量的组件，并在这些组件的基础上构建了 SSM 框架，它们为程序员进行应用程序的开发带来了极大的方便。

3. MVC 三层结构应用程序开发的优缺点

一个典型的应用软件常包括展现给用户界面的编码、业务处理模块编码、数据访问处理编码几个部分。如果将这些部分放在一起编程，则应用软件内部各元素耦合性非常高；现在人们常将它们分开来开发，然后将它们组装为一个整体。

作为一种设计模式，MVC 优缺点并存。

MVC 有以下优点：耦合性低、重用性高、利于分工开发、可维护性高、有利于软件工程化管理等。

1）耦合性低

MVC 程序中由于视图层和业务层分离，所以在更改视图层代码后不必重新编译模型层和控制器层代码。同样，由于模型层与控制器层和视图层相分离，一个应用程序的业务流程或者业务规则发生改变后，只需要改动 MVC 的模型层即可。

例如，把数据库从 MySQL 移植到 Oracle，只需改变模型即可。由于 MVC 的三个部件相互独立，改变其中一个不会影响其他两个，所以用这种模式设计的软件具有良好的松散耦合性。

2）重用性高

MVC 允许使用各种不同样式的视图来访问同一个服务器的代码，因为多个视图能共享一个模型。例如，用户可以通过计算机订购某件产品，也可以通过手机订购某件产品。虽然订购的方式不一样，但处理订购商品的方式是一样的，所以对应的模型可以是一样的。

由于对模型返回的数据没有进行格式化，所以同样的构件能被不同的界面使用。这些视图只需要改变视图层的实现方式，无需对控制层和模型层做任何改变，所以可以最大化地重用代码。

3）利于分工开发

使用 MVC 利于团队协作开发，从而大幅度缩短开发时间。它使程序员（Java 开发人员）集中精力于业务逻辑，界面程序员（HTML 和 JSP 开发人员、界面美工人员）集中精力于表现形式。

4）可维护性高

由于 MVC 的软件开发具有松散耦合性，它将视图层和业务逻辑层分离，因此应用程序更易于维护和修改。

5）有利于软件工程化管理

由于不同层各司其职，每一层不同的应用具有某些相同的特征，这样就可以对程序进行工程化、工具化管理。控制器可用来连接不同的模型和视图去完成用户的需求，从而为构造应用程序提供强有力的手段。给定一些可重用的模型和视图，控制器可以根据用户的需求选择模型进行处理，然后选择视图将处理结果显示给用户。

另外，由于 MVC 内部原理比较复杂，理解起来并不容易。所以，在使用 MVC 时需要精心地计划，需花费一定时间去思考。MVC 有调试较困难、不利于中小型软件开发、增加系统结构和实现的复杂性、视图与控制器耦合度过强、视图对模型数据的访问效率低的缺点。

1）调试较困难

模型和视图分离给调试应用程序带来了一定的困难，所以每个部件在使用之前都需要经过彻底的测试。

2）不利于中小型软件开发

花费大量时间将 MVC 应用到规模并不很大的应用程序，在工作量、成本、时间等方面常常得不偿失，所以对中小型软件的开发，可不选择 MVC 模式。

3）增加系统结构和实现的复杂性

对于简单界面的开发，也需严格遵循 MVC，使模型、视图与控制器分离，这会增加结构的复杂性，并可能产生过多的更新操作，降低运行效率。

4）视图与控制器耦合度过强

视图与控制器虽相互分离，但却是联系紧密的部件，没有控制器的存在，视图的应用是很有限的，反之亦然，这样就妨碍了它们的独立重用。

5）视图对模型数据的访问效率低

由于模型接口的不同，视图可能需要多次调用才能获得足够的显示数据。对未变化数据的不必要的频繁访问，也将损害操作性能。

MVC 设计模式为某一类问题提供了通用的解决方案，同时优化了代码，从而使代码更容易被人理解，提高了代码的复用性，并保证了代码的可靠性。

1.2　IntelliJ IDEA 平台简介

本书选用商业软件 IntelliJ IDEA（下文简称 IDEA）作为开发平台。IDEA 可用于开发 Java 程序，在业界被公认为较好的 Java 程序开发平台之一，其在智能编程、代码自动提示、重构、Java EE 支持、Ant、JUnit、CVS 整合、代码审查、GUI（图形用户界面）设计等方面的功能可以说是异常强大的。IDEA 有即刻完成的特性，例如，只需要输入单词的首字母，IDEA 就会立即给出最相关、最适合此处的代码编辑的选项供你选择。

1.2.1　IntelliJ IDEA 配置

与大多数 IDE（开发平台）一样，IDEA 也没有集成 JDK、Tomcat 等软件。因此，在使用 IDEA 之前首先要安装 JDK，并配置环境变量。要在 IDEA 中编写 Java/Java EE Web 程序，需要对其进行 JDK 与 Tomcat 的配置。

1.3 节将介绍在 IDEA 平台上对 JDK、Tomcat 的配置操作，然后通过编写并运行案例代码，测试其是否配置成功。

本书案例所使用的软件如下：

（1）JDK 1.8。

（2）Tomcat 9。

（3）IDEA 2016。

（4）MySQL 5.7。

1.2.2　IntelliJ IDEA 界面说明和快捷键

1. IDEA 界面说明

IDEA 主界面如图 1-2 所示。

同其他 IDE 工具一样，IDEA 主界面有菜单栏、工具栏、导航条、状态栏等。

IDEA 主界面中最上面是菜单栏，用户可以通过菜单栏中的菜单执行相应的命令；也可以通过弹出式菜单更方便地执行命令，如对源文件、类的操作等。主界面最下面是状态栏，状态栏可显示项目名称、整个 IDE 的状态，并显示各种警告信息等。下面对 IDEA 主界面其他部分进行简要说明。

图 1-2 中①是工具栏，工具栏包含了与项目相关的通过导航条大部分命令。②是导航条，导航条有助于了解当前页面在该项目中的位置。通过导航条可方便地对文件进行浏览或对文件进行编辑。

图 1-2 中③都是工具窗口。工具窗口有多个，它们执行不同的功能，如通过项目和文件结构导航，查看搜索和检查结果，运行、调试和测试应用程序，通过控制台交互等。

图 1-2 中④是编辑器，可以创建和修改代码。

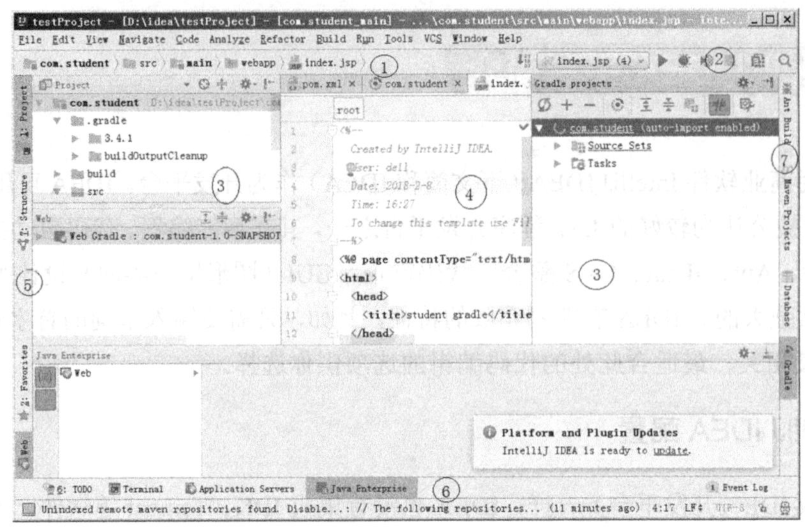

图 1-2　IDEA 主界面

图 1-2 中⑤、⑥、⑦分别为左侧边栏、下侧边栏、右侧边栏。这些侧边栏，不但可以显示当前编辑器中打开的文件的断点，而且提供了一种便捷的方式，可分类导航到各个不同的功能窗口。当开发者不需要该功能窗口时，又可以将其关闭。

这些功能条不仅仅是流动条，它们还能不断地监视代码的质量，始终显示代码分析的结果：错误、警告等。如果搜索已打开文件中的代码，右侧边栏上会以绿色标出搜索到的代码位置，单击绿色的部分就可以快速定位目标。

2. IDEA 常用快捷键

IDEA 内置了很多快捷键，以方便开发人员使用，这也是 IDEA 受欢迎的原因之一。下面介绍在 Windows 系统下比较常用的快捷键（见表 1-1），掌握这些快捷键可以大大提升编码效率。

表 1-1　IDEA 常用快捷键

快　捷　键	功　　能
Alt+Enter	导入包，自动修正
Ctrl+N	查找类
Ctrl+Shift+N	查找文件
Ctrl+Alt+L	格式化代码
Ctrl+Alt+O	优化导入的类和包
Alt+Insert	生成代码（如 get、set 方法，构造函数等）
Ctrl+R	替换文本
Ctrl+F	查找文本
Ctrl+Shift+Space	自动补全代码
Ctrl+Space	代码提示
Ctrl+Alt+Space	类名或接口名提示
Ctrl+P	方法参数提示
Ctrl+Shift+Alt+N	查找类中的方法或变量
Alt+Shift+C	对比最近修改的代码
Shift+F6	重构或重命名

续表

快　捷　键	功　　能
Ctrl+X	删除行
Ctrl+D	复制行
Ctrl+/ 或 Ctrl+Shift+/	注释（// 或/*...*/ ）
Ctrl+E	打开最近打开的文件
Ctrl+H	显示类结构图
Ctrl+Q	显示注释文档
Alt+F1	查找代码所在位置
Alt+1	快速打开或隐藏工程面板
Ctrl+Alt+left/right	返回至上次浏览的位置
Alt+left/right	切换代码视图
Alt+Up/Down	在方法间快速移动
Ctrl+Shift+Up/Down	代码向上/下移动
F2 或 Shift+F2	高亮错误或警告快速定位
Tab	代码标签输入完成后，按该键，生成代码
Ctrl+Shift+F7	选中文本，按该快捷键，高亮显示所有该文本，按 Esc 键高亮消失
Ctrl+W	选中代码，连续按该快捷键，会有其他效果
Alt+F3	选中文本，按该快捷键，逐个往下查找相同文本，并高亮显示
Ctrl+Up/Down	光标跳转到第一行或最后一行
Ctrl+B	快速打开光标处的类或方法

1.2.3　IntelliJ IDEA 简单断点调试

Debug（调试）模式用来追踪代码的运行流程和异常。当程序在运行过程中出现异常，启用 Debug 模式可以确定异常发生的位置，以及了解运行过程中参数的变化。我们也可以启用 Debug 模式来跟踪代码的运行流程去学习三个框架的源码。我们使用图 1-3 简单讲解在 IDEA 中如何进行断点调试。

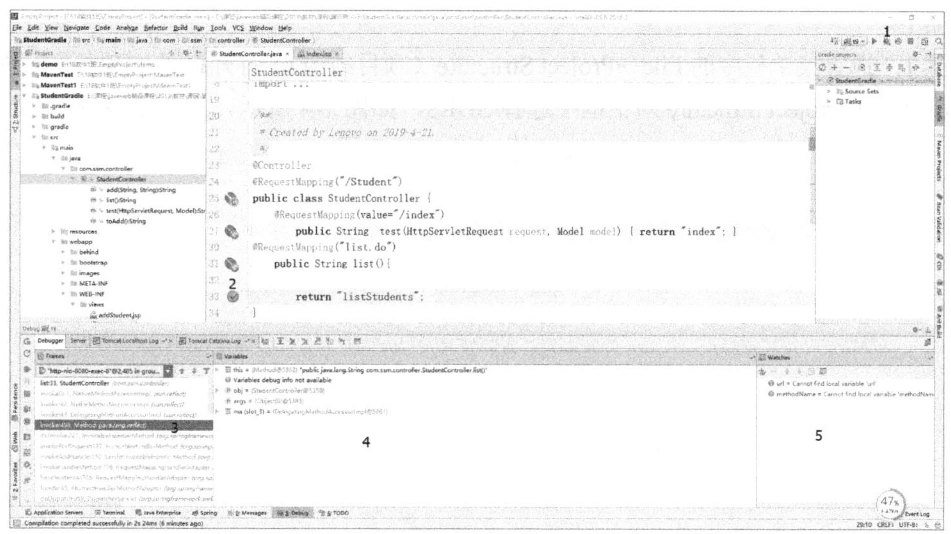

图 1-3　IDEA 断点调试界面

标号 1——以 Debug 模式启动 Tomcat 服务。标号 1 左边的按钮▶表示以 Run 模式正常启动。在开发中，一般直接启动 Debug 模式，方便随时调试代码。

标号 2——设置断点。在左边行号栏单击，或者利用快捷键 Ctrl+F8 插入/取消断点，断点行的颜色可自行设置。

标号 3——Debug 窗口。访问请求到达第一个断点后，会自动激活 Debug 窗口。如果没有自动激活，可以去设置（选择菜单栏中的 File→Settings）。

标号 4——Variables（变量区）。在变量区可以查看当前断点之前的当前方法内的变量。

标号 5——实时的变量跟踪。参数所在行后面会显示当前变量的值。如果光标悬停到参数上，也会显示当前变量信息。

1.3 IntelliJ IDEA 平台准备及简单 Web 项目创建

下面介绍在 IntelliJ IDEA 2016 中集成 JDK 1.8、Tomcat 9 的过程，并通过配置好的开发平台创建一个简单的 Web 项目，然后在 Tomcat 下运行以测试其是否配置成功。

1.3.1 在 IntelliJ IDEA 中配置 JDK 并测试

现在大多数 IDE 都没有集成 JDK，IDEA 也一样，在使用 IDEA 之前首先要安装 JDK 1.8，并且配置环境变量。JDK 的安装按导航一步步操作即可。

1. 在 IDEA 中进行 JDK 配置操作

与其他 IDE 的不同之处在于，IDEA 不会自动匹配 JDK。如果在 IDEA 中没有配置 JDK，那么运行程序时就会报错。

接下来开始配置 JDK，IDEA 中可以同时有多个版本的 JDK，并且需要开发者手动配置项目所使用的 JDK 版本。也就是说，对不同的项目，我们可以选择使用不同版本的 JDK。（同样，在 IDEA 中 Tomcat 也可以有不同的版本），此处选择 JDK 1.8。在 IDEA 中配置 JDK 1.8 的操作步骤如下。

步骤 1．选择菜单栏中的 File→Project Structure（项目结构），打开 Project Structure 对话框。

步骤 2．在 Project Structure 对话框中选择 SDKs，如图 1-4 所示，进入 SDK 配置界面。

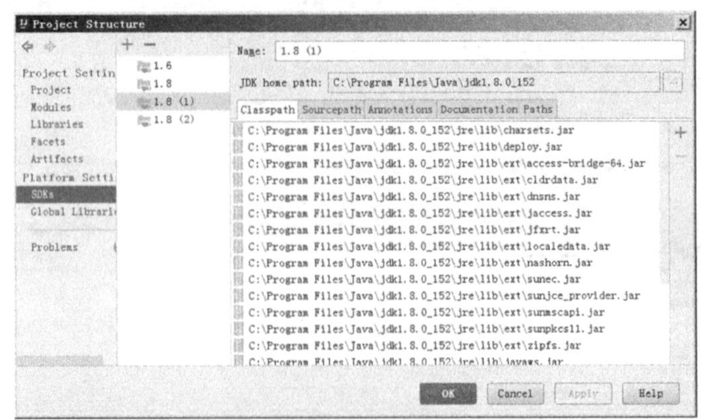

图 1-4　选择 SDKs 列表项

这样就可以在中间的列表框中选择自己需要的 JDK 版本了。

步骤 3．若中间的列表框中没有选项，则单击“+”创建一个 JDK 列表项。

步骤 4．在中间的列表框中选择 1.8（1）选项，单击 JDK home path 浏览按钮，指定 JDK 安装路径并保存。

也可以在 Project Structure 对话框中选择 Project →Project SDK（见图 1-5）。要添加新的 JDK 版本，只需要单击 New 按钮选择对应的路径就可以了。还有另一种方式，可以在创建项目后设置 SDK。选择菜单栏中的 File→Project Structure，弹出如图 1-5 所示对话框，选择 Project→Project SDK，在下拉列表里选择安装好的 JDK 1.8。

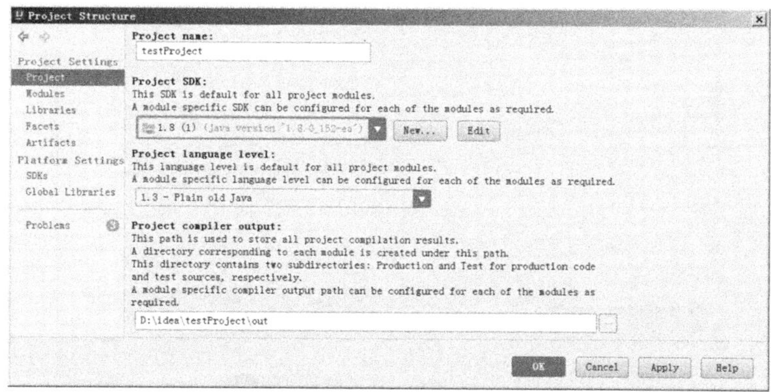

图 1-5　JDK 另一种配置方式

上述操作完成后，再创建 Java 项目，就会出现 JDK1.8 的选项配置。这就表明可以编写与运行 Java 程序了。

2．编写简单的 Java 程序来测试 JDK 配置是否成功

下面通过一个简单的例子，即在 IDEA 中创建、编辑与运行一个 Java 程序，测试上述 JDK 配置是否成功。

步骤 1．选择菜单栏中的 File→New→Project，创建一个 Project（项目），如图 1-6 所示。

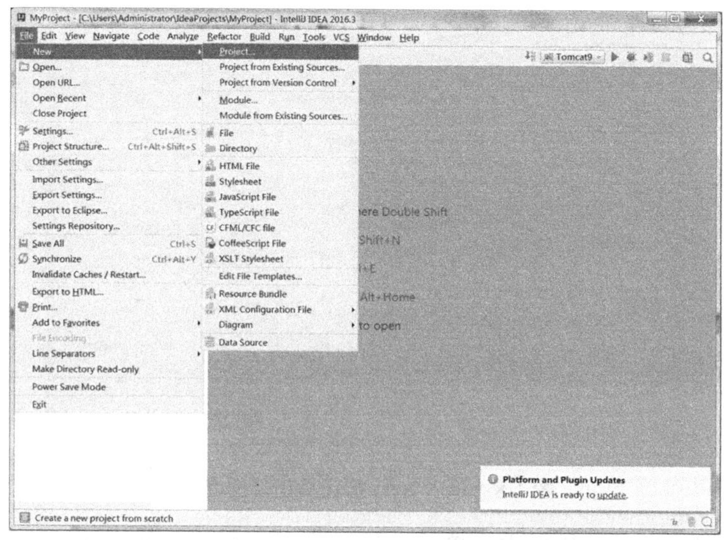

图 1-6　创建 Project 的菜单

步骤 2．出现如图 1-7 所示界面，在该界面中勾选 Create project from template（从模板中创建项目）复选框，并选择 Java Hello World 选项（见图 1-7）。

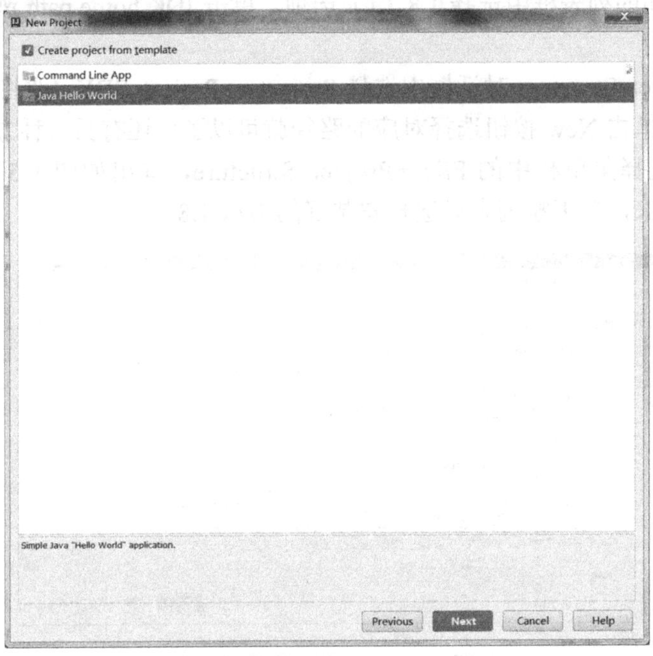

图 1-7　勾选 Create project from template 复选框

步骤 3．单击 Next 按钮，弹出如图 1-8 所示界面。

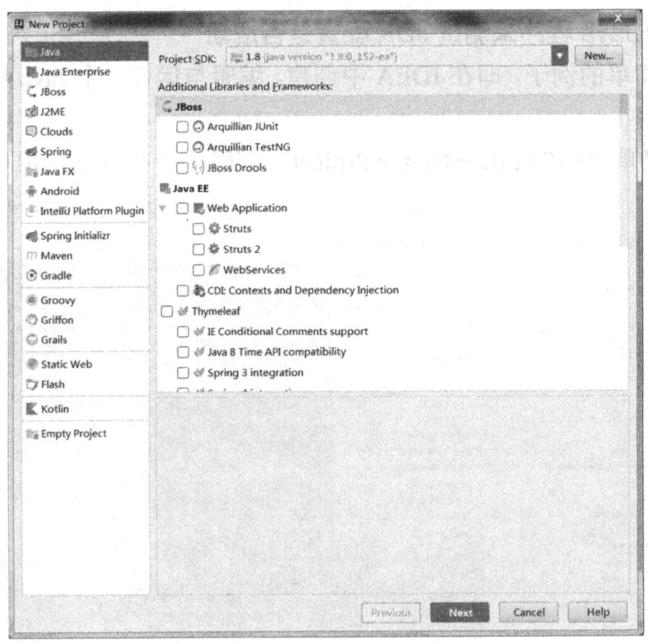

图 1-8　单击 Next 按钮后出现的界面

步骤 4．在图 1-8 所示的界面中直接单击 Next 按钮，弹出如图 1-9 所示界面。在该界面中给项目取名。

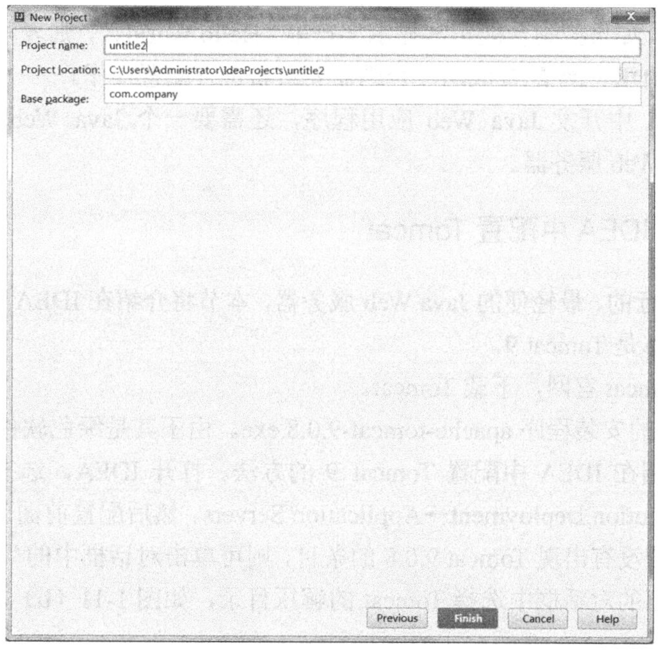

图 1-9　给 Project 命名

步骤 5．在 Project name 输入框中给该项目取一个名称，其他输入框可用默认值。最后单击 Finish 按钮，然后出现图 1-10 所示界面。

在图 1-10 中出现了一个类 Main.java，该类的内容是在控制台显示一个字符串“Hello World!”。

由于是做简单测试，所以该程序可不修改。

图 1-10 右上部工具栏中有一个 Run（运行）按钮▶，单击该按钮就可以运行该程序。

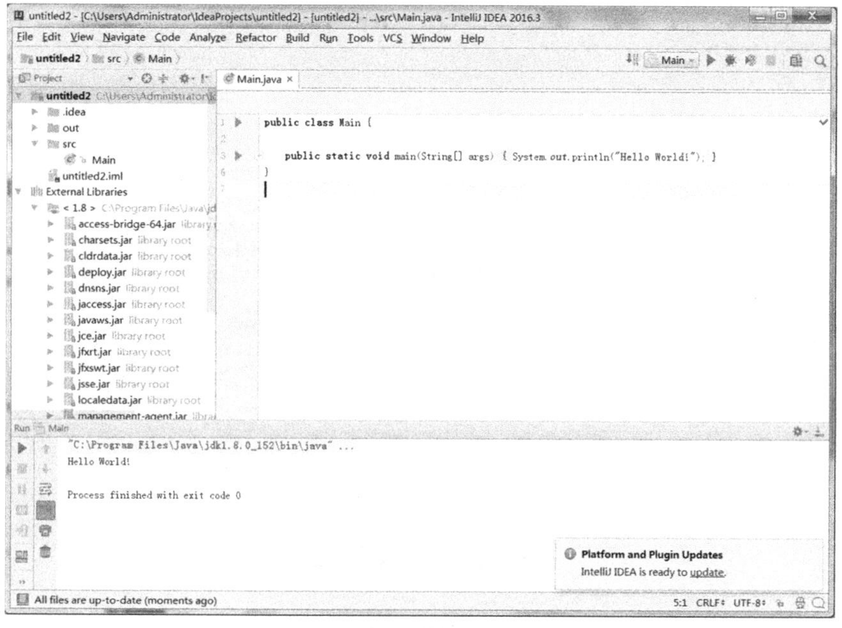

图 1-10　Java 程序编辑及运行结果

单击 Run 按钮，如果控制台能正常显示字符串“Hello World!”，则表明 IDEA 中支持 Java 程序开发与运行的 JDK 已经配置成功，否则需要重新进行配置操作。

但若要在 IDEA 中开发 Java Web 应用程序，还需要一个 Java Web 服务器。本书选择 Tomcat 9 作为 Java Web 服务器。

1.3.2 在 IntelliJ IDEA 中配置 Tomcat

Tomcat 是最流行的、最轻便的 Java Web 服务器，本节将介绍在 IDEA 中集成 Tomcat 的操作。这里选择的版本是 Tomcat 9。

首先，访问 Tomcat 官网，下载 Tomcat。

下载 Tomcat 9 的安装程序 apache-tomcat-9.0.8.exe。由于其是绿色软件，安装比较简单，解压即可。下面介绍在 IDEA 中配置 Tomcat 9 的方法。打开 IDEA，选择菜单栏中的 File→Settings→Build,Execution,Deployment→Application Servers。然后配置前面创建的 Tomcat 9，如图 1-11（a）所示。若没有出现 Tomcat 9.0.8 的条目，则可单击对话框中的“+”号，选择 Tomcat Server 选项，在弹出的对话框中选择 Tomcat 的解压目录，如图 1-11（b）所示。

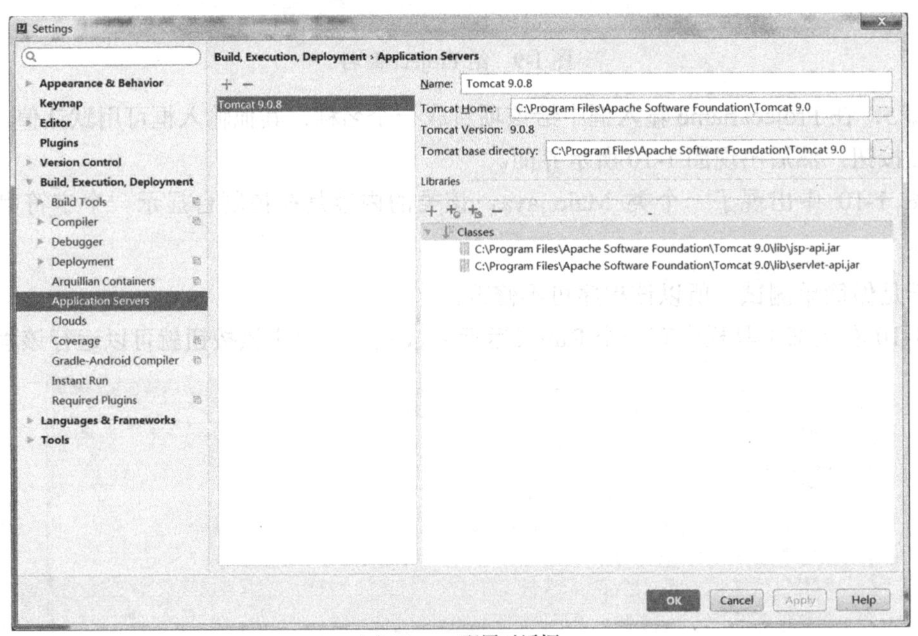

（a）Tomcat 配置对话框

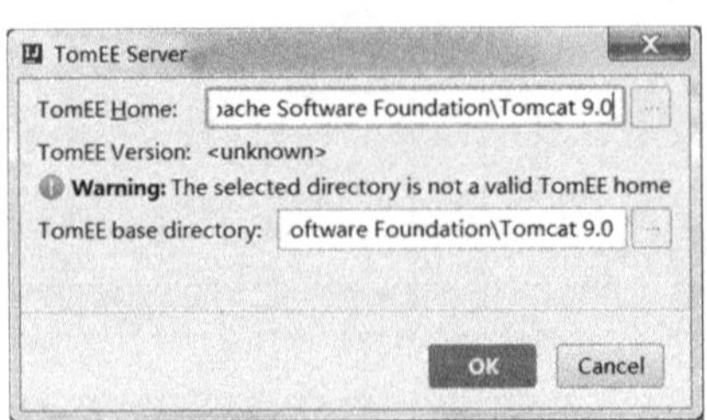

（b）选择 Tomcat 解压目录

图 1-11　Tomcat 配置界面

图 1-11（a）是 Tomcat 配置对话框，用于配置 Tomcat 服务器名称、版本号及安装目录。通过界面中的“+”“-”号，可以增加或删除所配置的服务器。图 1-11（b）所示就是通过单击“+”号，对该 Tomcat 服务器的安装目录进行选择的对话框。这里选择的是上面介绍的 Tomcat 9 的安装目录。

注 意 若选择目录后，提示 Tomcat 版本未知，并且目录显示红色，则是因为 Tomcat 执行权限的问题。解决的方法：进入 Tomcat 的安装目录，为 apache-tomcat-9.0.8 整个文件夹及子文件赋予读写权限。

在 IDEA 中配置好 Tomcat 服务器后，就可以在其中开发 Web 项目并运行该项目了。

1.3.3 任务：创建简单的 Web 项目并在 Tomcat 下运行

配置好 JDK 及 Tomcat 的 IDEA 开发平台后，就可以创建 Web 项目，并在 Tomcat 下运行以检验 IDEA 开发平台是否配置成功。

在 IDEA 中，要分清两个概念：Project 和 Module。其中，Project 是一个完整的项目，而 Module 是项目中的一个小模块，而且一个 Module 可以包含多个 Module。下面介绍在 IDEA 中创建 Web 项目的步骤。

1. 创建一个 Project

在 IDEA 中选择菜单栏中的 File→New→Project，创建一个 Project，弹出如图 1-12 所示的对话框。

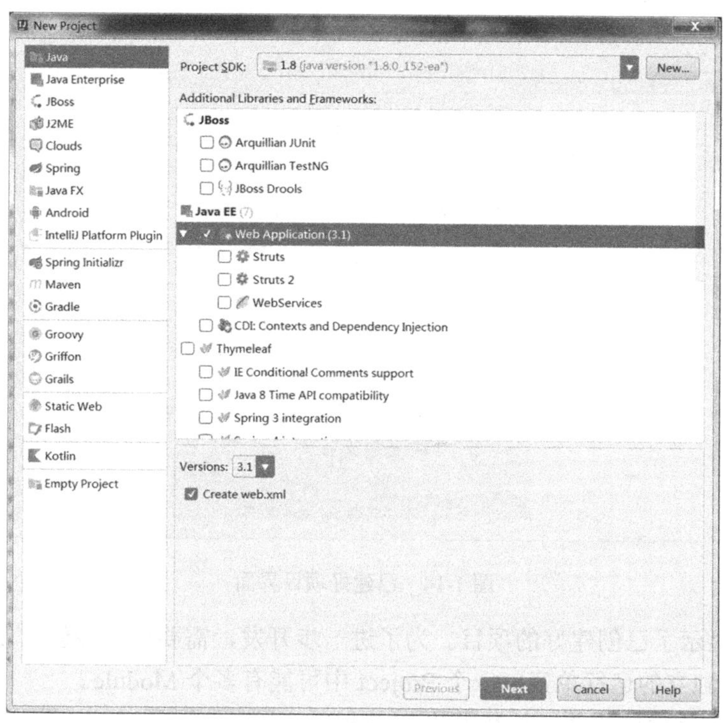

图 1-12 New Project 对话框

勾选 Web Application 复选框和 Create web.xml 复选框，单击 Next 按钮，弹出如图 1-13

所示的对话框。在该对话框中给 Project 取一个名称：MyProject。单击 Finish 按钮，进入如图 1-14 所示界面。

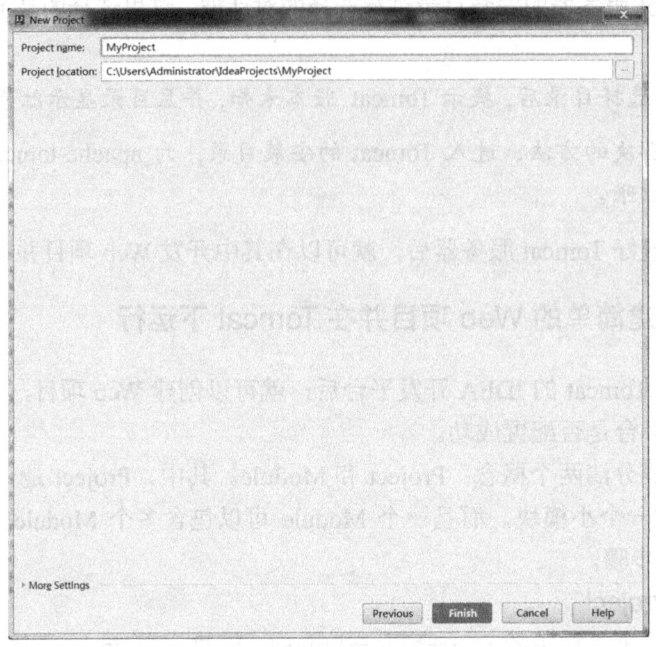

图 1-13　给 Project 命名

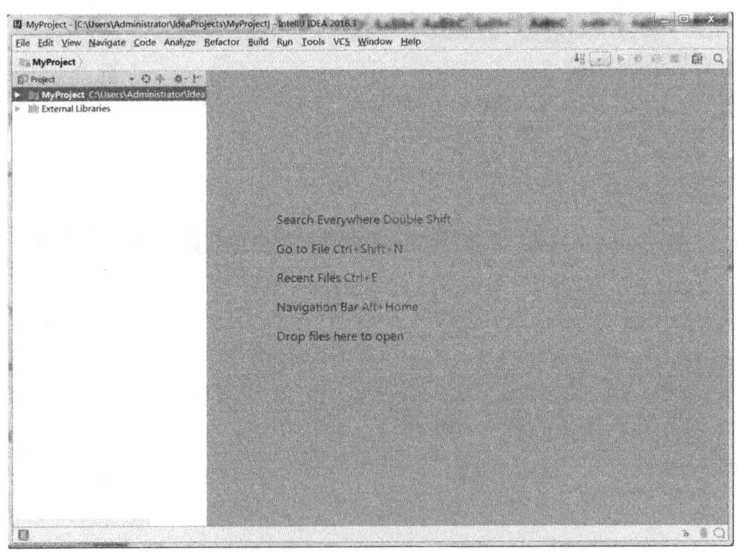

图 1-14　已建好项目界面

图 1-14 中显示了已创建好的项目。为了进一步开发，需要进一步创建 Module。Module 是 Project 中相对独立的运行单元，一个 Project 中可能有多个 Module。

下面介绍如何创建模块并运行其中的 Web 程序。

2．创建一个 Module

在图 1-14 所示的界面中，选中项目名称 MyProject，然后右击，在弹出的快捷菜单中选择

New→Module，弹出如图 1-15 所示对话框。

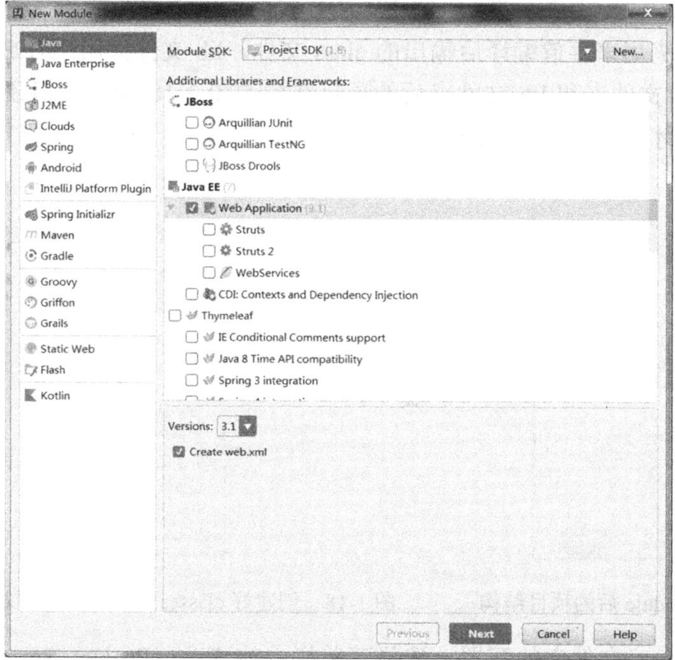

图 1-15 New Module 对话框

在该对话框中勾选 Web Application 复选框，且确认已勾选 Create web.xml 复选框，然后单击 Next 按钮，弹出如图 1-16 所示对话框。

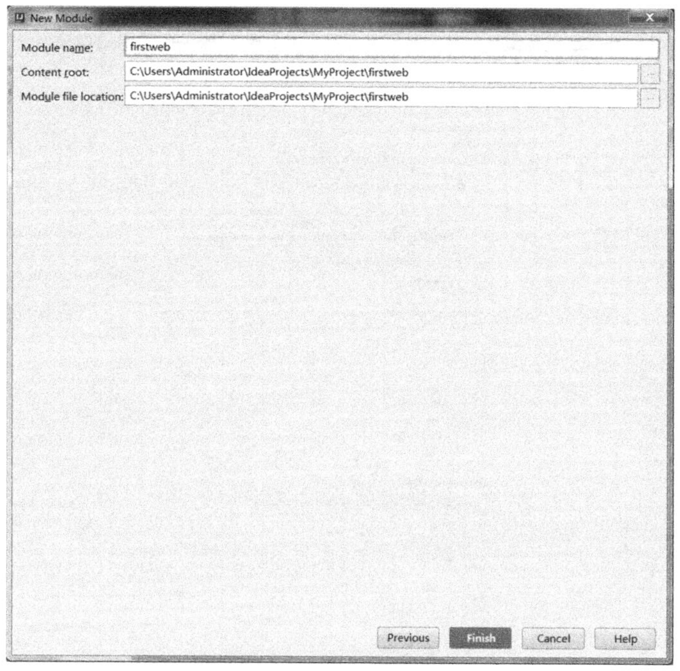

图 1-16 给 Module 命名

在该对话框中，给 Module 命名：firstweb。然后单击 Finish 按钮，则弹出如图 1-17 所示的创建好 Module 后的项目结构。

下面还要在 firstweb 模块下的 web\WEB-INF 文件夹下创建两个子文件夹：classes 和 lib。其中，classes 文件夹用来存放编译后输出的 class 文件，lib 文件夹用于存放第三方 jar 包。

创建好 classes 文件夹和 lib 文件夹后的项目结构如图 1-18 所示。但还需要对这两个文件夹配置相应的路径才算完成 Module 的创建操作。

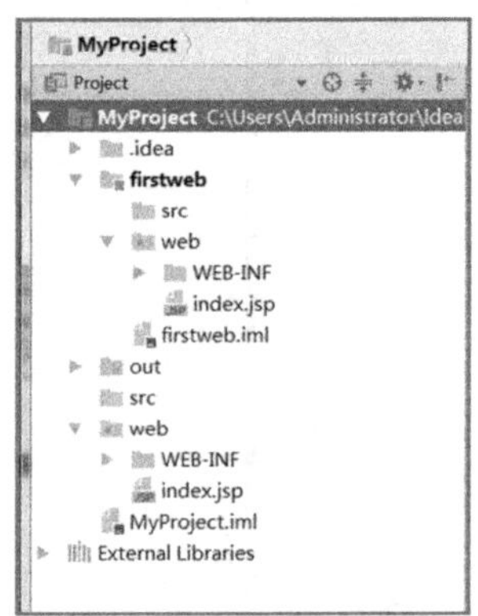

图 1-17 创建好 Module 后的项目结构

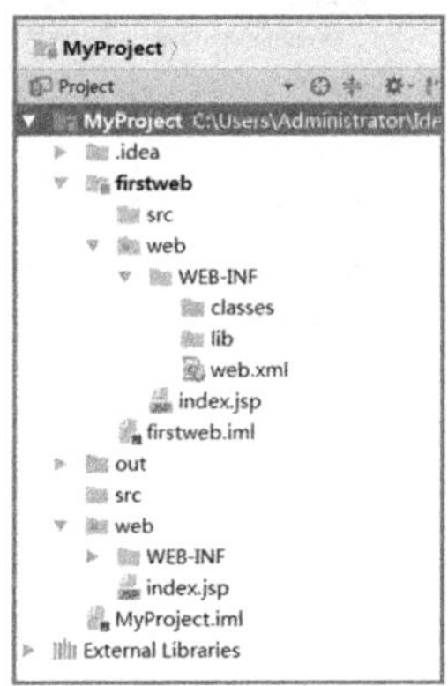

图 1-18 创建好 classes 文件夹和 lib 文件夹后的项目结构

首先介绍 classes 文件夹的配置。选择菜单栏中的 File→Project Structure，弹出 Project Structure 对话框，选择 Modules→firstweb。

然后选择 Paths 选项卡，则弹出如图 1-19 所示界面。并选择 Use module compile output path（使用模块的编译输出路径）单选按钮，Output path 和 Test output path 都选择刚刚创建的 classes 文件夹。

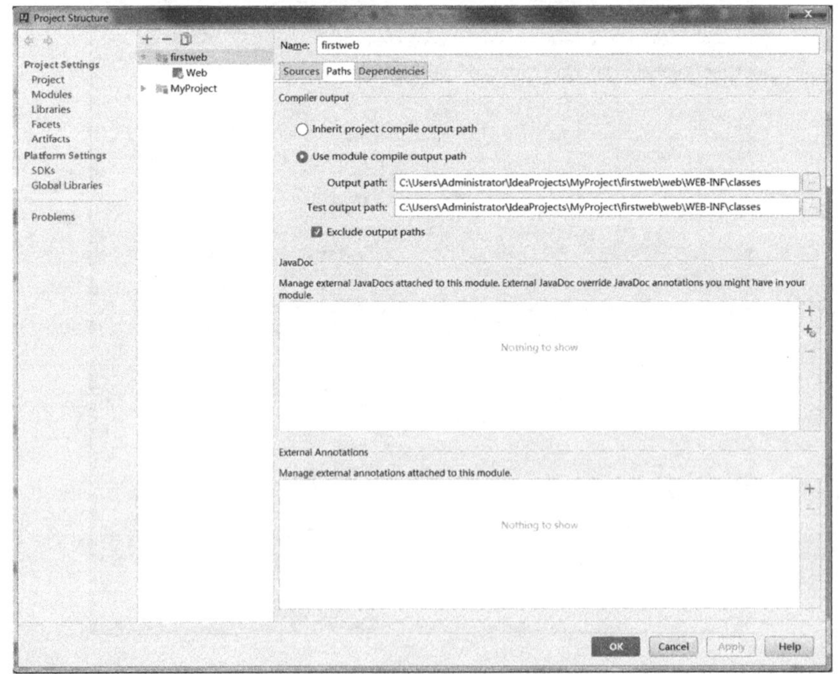

图 1-19 配置 classes 文件夹的路径

下面还要配置 lib 文件夹，其操作如下。

如图 1-20 所示，选择 Dependencies 选项卡，将 Module SDK 设置为 Project SDK（1.8），再单击右边的“+”号，选择 1 JARs or directories 选项。

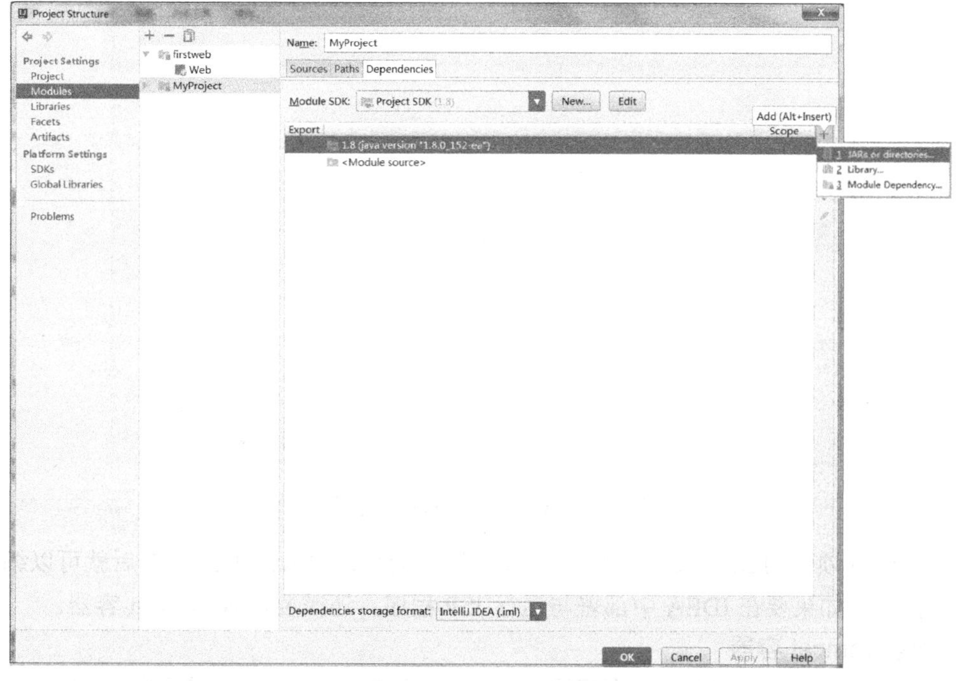

图 1-20　配置 lib 文件夹的路径

在弹出的对话框中选择刚刚创建的 lib 文件夹，如图 1-21 所示。

然后单击 OK 按钮。在弹出的如图 1-22 所示的对话框中选择 Jar Directory 选项，接着单击 OK 按钮。

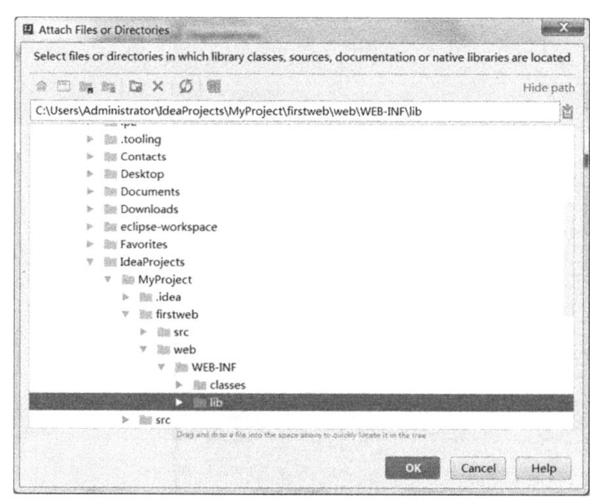

图 1-21　选择 lib 文件夹

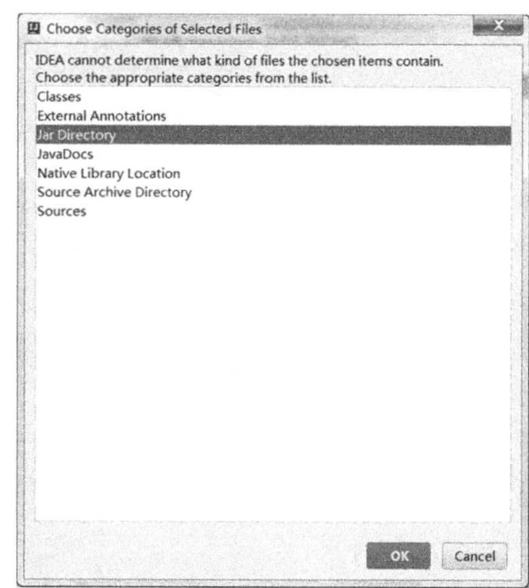

图 1-22　选择 Jar Directory

lib 文件夹配置成功的界面如图 1-23 所示。

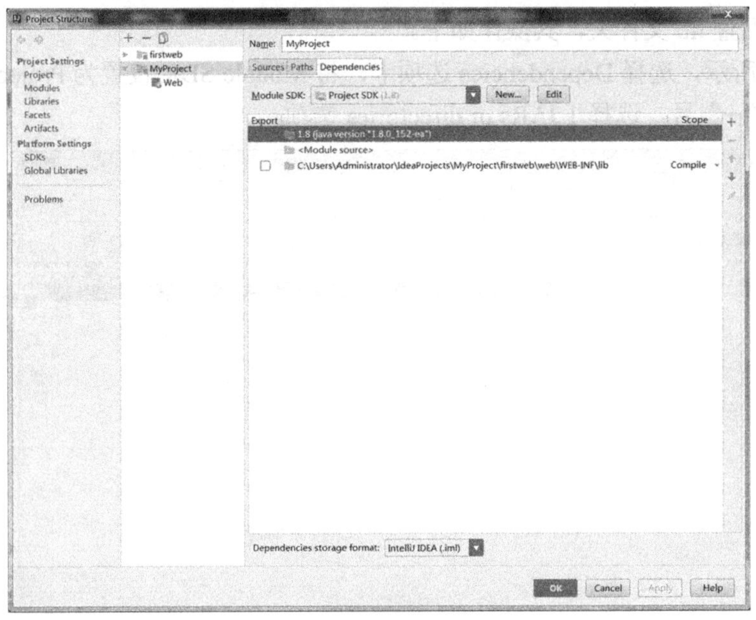

图 1-23　lib 文件夹配置成功的界面

至此，Web 项目的 Module 创建成功，其项目结构可以参考图 1-24。然后就可以编写 JSP 程序了。但是，如果要在 IDEA 中部署与运行 JSP 程序，还需要配置 Tomcat 容器。

3．配置 Tomcat 容器

前面已经介绍了如何在 IDEA 中配置 Tomcat 服务器。但是，要想将刚创建的 Web 项目中的 JSP 程序部署到 Tomcat 中并运行，还需要为该项目配置 Tomcat 服务器。下面就介绍 Tomcat 服务器的配置。

配置 Tomcat 服务器的操作步骤如下。

步骤 1．选择菜单栏中的 Run→Edit Configurations（编辑配置），如图 1-24 所示。

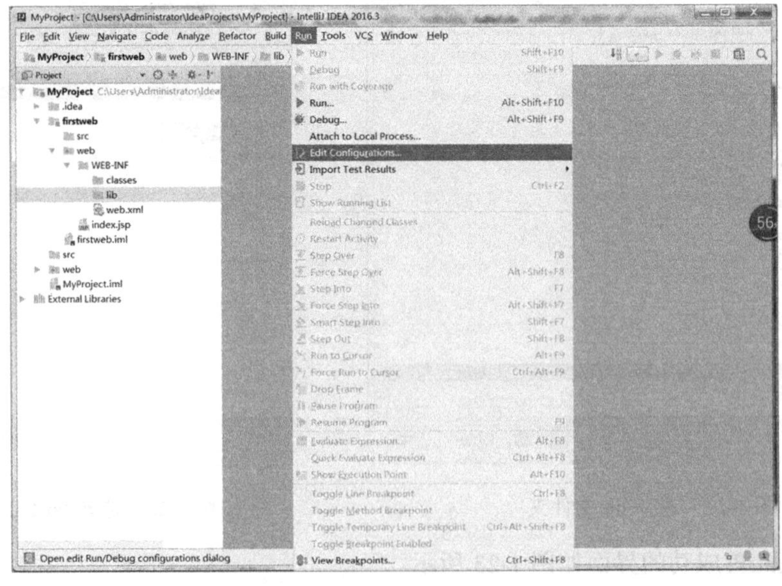

图 1-24　配置 Tomcat 服务器的菜单

步骤 2. 在出现的对话框中，单击“+”号，选择 Tomcat Server（Tomcat 服务器）→Local（本地），如图 1-25 所示。

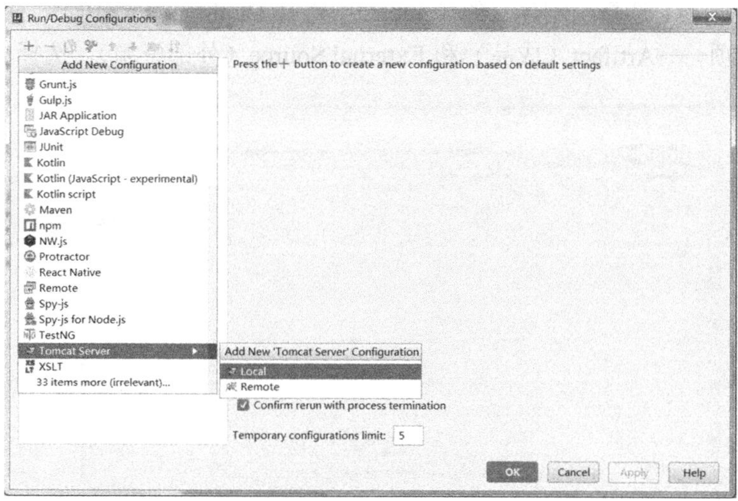

图 1-25　选择 Tomcat Server→Local

步骤 3. 单击 OK 按钮，在弹出的对话框中的 Name 输入框中输入新的 Tomcat 服务器名，单击 Application server 后面的 Configure 按钮，如图 1-26 所示。弹出 Tomcat Server 对话框，选择本地安装的 Tomcat 目录。

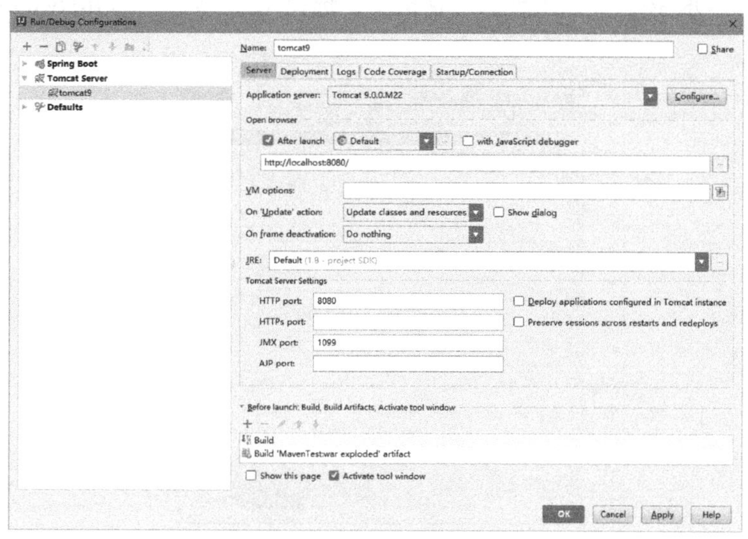

图 1-26　配置 Tomcat 对话框

在如图 1-26 所示的对话框中，在 Server 选项卡中，取消勾选 After launch 复选框。On'Update'action 选项表示当用户主动执行更新的时候更新，这里将其设置为 Update classes and resources，即用户更新类和资源时更新部署项目；On frame deactivation 选项表示在编辑窗口失去焦点的时候更新，根据需要设置该选项即可。设置 HTTP port（默认值为 8080）和 JMX port（默认值为 1099）（如果在运行时出现端口已被占用的异常，则修改为其他未被占用的端口号）。最后，单击 Apply 按钮，再单击 OK 按钮关闭对话框，至此 Tomcat 服务器配置完成。

4．在 Tomcat 中部署并运行项目

步骤 1．选择菜单栏中的 Run→Edit Configurations，弹出 Run/Debug Configurations 对话框，如图 1-27 所示。选择刚刚建立的 Tomcat 9，选择 Deployment 选项卡，单击右边的“+”号，弹出两个选项——Artifact（成品）和 External Source（外部资源），选择 Artifact 选项。

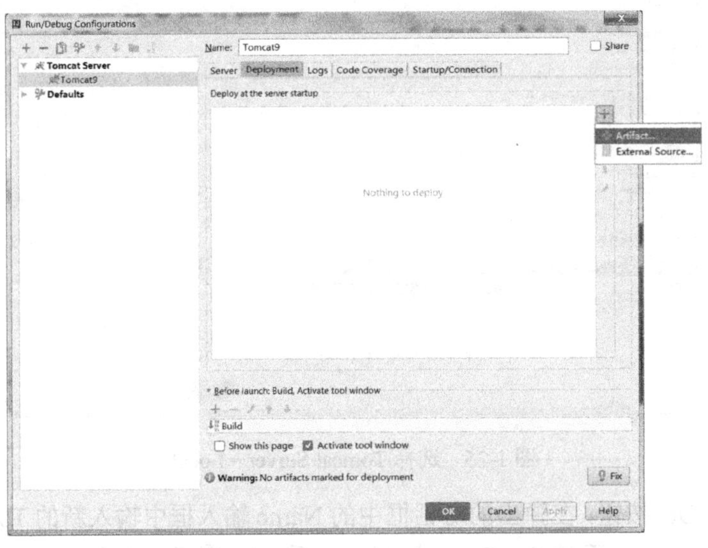

图 1-27　Run/Debug Configurations 对话框

步骤 2．单击 OK 按钮，在出现的对话框中，选择 Web 项目（如 firstweb:war exploded），如图 1-28 所示。

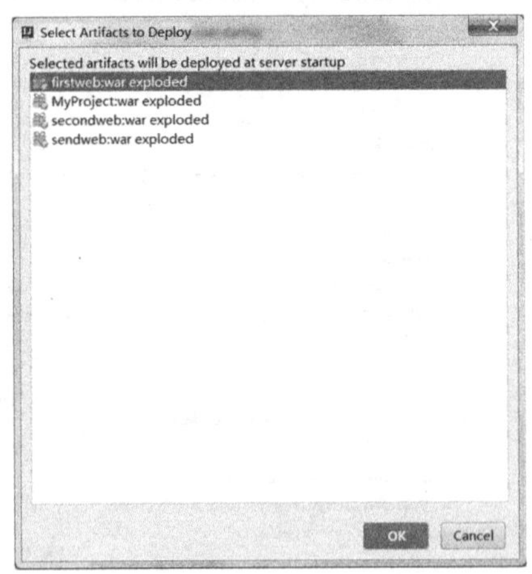

图 1-28　选择部署的项目

步骤 3．单击 OK 按钮，在出现的对话框中显示了已部署的 firstweb 模块。再在 Application context 下拉列表中选择“/hello”（也可以不选），并单击 OK 按钮完成操作。部署 firstweb 成功的界面如图 1-29 所示。

提示：部署 Artifact 时有两种选择：一种是“项目名:war exploded”类型的，表示项目展开后部署，相当于将资源文件展开后进行部署。另一种是“项目名:war”类型的，称为发布模式，表示先打成 war 包，再部署。如果需要热部署，即时显现项目编辑后的效果，则应选择“项

目名:war exploded”类型。

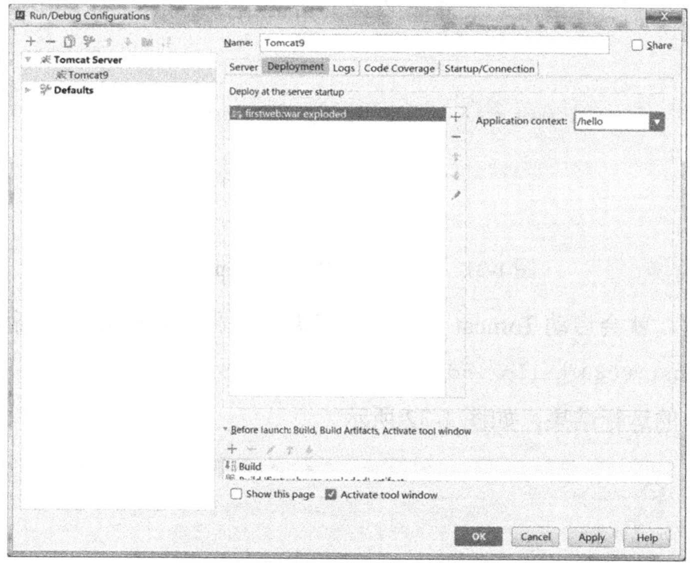

图 1-29　部署 firstweb 成功的界面

至此，在 Tomcat 9 容器中已经部署成功 firstweb 模块，下面就可以在其中编写 JSP 程序并运行了。

5．编辑 index.jsp 文件并运行

firstweb 模块中已经有默认主页程序 index.jsp，我们编辑并运行它（当然，也可以新建一个 JSP 程序）。在如图 1-30 所示窗口中双击 index.jsp 文件，出现其编辑界面，如图 1-30 所示，在其中输入：

```
HelloWorld!
```

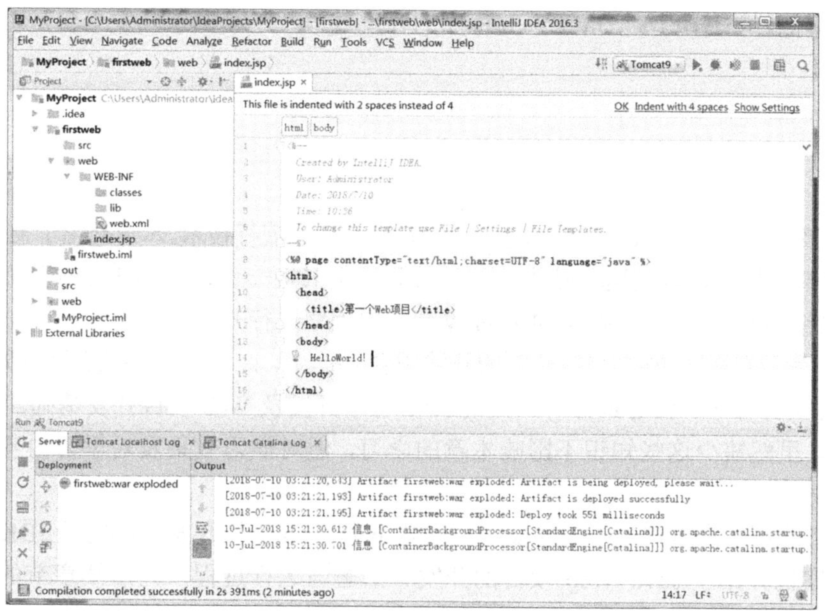

图 1-30　编辑 index.jsp 动态页面程序

编辑好 index.jsp 文件并保存后，就可以运行了。在运行前，先要启动 Tomcat 服务器。在配置好 Tomcat 9 容器后，在 IDEA 主界面的右上角就出现了 Tomcat 的 Run 与 Stop 按钮，如图 1-31 所示。

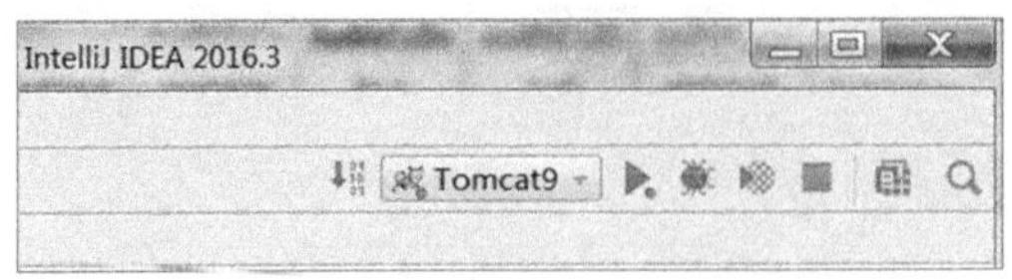

图 1-31　Tomcat 的 Run 与 Stop 按钮

单击 Run 按钮，就会启动 Tomcat 服务器。打开浏览器，在地址栏中输入如下地址：

```
http://localhost:8080/hello/index.jsp
```

就能看到 index.jsp 的运行结果，如图 1-32 所示。

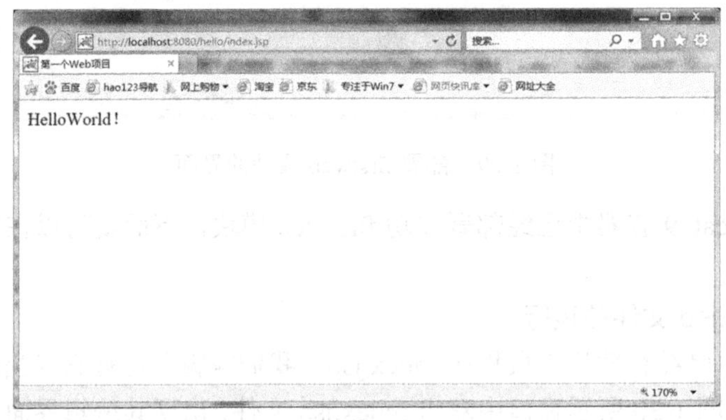

图 1-32　index.jsp 运行结果

至此，第一个 Web 项目运行结束。

通过该案例，不仅介绍了 Java Web 程序的开发过程，还测试了开发平台是否配置成功。

小　结

Java EE 及 SSM 是基于 Java 平台的适合大型软件开发的流行技术。本章介绍了 MVC、Java EE、SSM 框架等基本概念，以及这些技术之间的关系。本书将一直采用 IDEA 作为开发平台，对 Java EE 体系结构及 SSM 框架技术的 Java Web 应用程序进行开发。

IDEA 作为一个用于 Java 程序开发的开发平台，提供了许多功能，帮助程序员提高了程序开发效率。学习本章前，读者最好有一定的 Java 程序开发及 Web 项目开发基础，并了解 MVC 设计模式等。虽然缺少这些知识不影响本章的学习，但对学习效果及对学习内容的领悟有一定影响。

本章着重介绍了在 IDEA 中如何安装和配置 Java Web 项目开发所必需的 JDK、Tomcat 软件，并通过案例介绍了 Java Web 项目的开发、部署与运行过程，为后续章节的学习打下了基础。

习　题

一、填空题

1．框架（Framework）就是已经开发好的一组_______，程序员可以利用它来快速地搭建自己的应用软件。

2．SSM 框架就是________、________、________三个框架的简称，它们是在________基础上创建的用于快速构建用户应用系统的软件框架集。

3．三层结构的应用程序包括：________、________、________。

4．MVC 模式的 M、V、C 分别表示___________、________、________。

5．IDEA 在业界被公认为是较好的________开发工具之一，尤其在智能编程、代码自动提示、重构、_______支持等方面的功能可以说是异常强大的。

二、简答题

1．什么是 Java EE？它有什么作用？

2．SSM 框架的含义是什么？并说明它与 Java EE 的关系。

3．三层结构的软件开发包括哪几层？MVC 设计模式的软件开发的含义是什么？它有什么优点？

4．分别阐述 Spring、Spring MVC、MyBatis 框架的基本内容与作用。

5．如何用 IDEA 进行 Java Web 应用程序开发？

综合实训

实训 1．自己动手搭建一个 IDEA 开发平台，并配置好 JDK、Tomcat。

实训 2．在已经搭建好的 IDEA 开发平台中，创建一个空的项目和 Java EE 模块。

第 2 章

项目构建工具 Gradle

本章学习目标

- 了解三大构建工具
- 熟悉 Gradle 的安装与配置
- 掌握在 IDEA 平台下创建 Gradle 项目的方法
- 掌握 Gradle 项目的配置和依赖管理的方法

本章介绍项目自动化构建工具 Gradle 的起源和使用。通过案例介绍如何在 IDEA 平台下一步步创建基于 Gradle 的模块，管理模块的依赖和运行任务，并演示如何将模块部署在 Tomcat 服务器上运行。本书以后章节的案例都是在本章案例基础上添砖加瓦，构建集成三大框架并实现基础业务功能的最终项目。

2.1 Gradle 简介和三大构建工具比较

2.1.1 Gradle 简介

Gradle 是基于 Apache Ant 和 Apache Maven 概念的项目自动化构建工具。它使用基于 Groovy 的特定领域语言（DSL）来声明项目设置，抛弃了基于 XML 的各种烦琐配置。

Gradle 这款基于 JVM 的构建工具，灵活好用、免费开源，支持多项目的构建，支持局部构建和多方式依赖管理：包括 Maven 远程仓库、Nexus 私服、Ivy 仓库及本地文件系统的 jars 包或者 dirs 目录文件。Gradle 与 Ant、Maven、Ivy 有良好的相容性。使用 Gradle 构建的项目可以在各种开发平台和服务器上轻松迁移，Gradle 适用于任何结构的项目，可以在同一个开发平台平行构建原项目和 Gradle 项目。

2.1.2 三大构建工具比较

Java 中主要有三大构建工具：Ant、Maven 和 Gradle。经过几年的发展，Ant 几乎销声匿迹，只能在 Eclipse 平台上看到它的身影。Maven 也日薄西山，一些企业因项目历史原因还沿用，而 Gradle 的发展则如日中天。

Maven 的主要特点是引入依赖管理系统，可实现多模块构建，具有一致的项目结构、一致的构建模型和插件机制。

Maven 为 Java 引入了一个新的依赖管理系统，可以用 groupId（组织名）、artifactId（成品名）、version（版本号）组成的 Coordination（坐标）唯一标识一个依赖。任何基于 Maven 构建的项目必须定义这 3 项属性，生成的包可以是 jar 包，也可以是 war 包或者 ear 包。一个典型的依赖引用如下：

```
<dependency>
 <groupId>org.Springframework</groupId>
 <artifactId>Spring-test</artifactId>
</dependency>
```

从上面代码可以看出当引用一个依赖时，version 可以省略，这样在获取依赖时会选择最新的版本。而存储这些组件的仓库有远程仓库和本地仓库之分。远程仓库可以使用世界公用的 Maven 中央仓库，也可以使用国内镜像阿里巴巴仓库，还可以使用 Apache Nexus 自建私有仓库；本地仓库则在本地计算机上。通过 Maven 安装目录下的 settings.xml 文件可以配置本地仓库的路径，以及采用的远程仓库的地址。在本书第 8 章，使用快速开发框架 Spring Boot 搭建项目时使用了 Maven，并详细讲解了 Maven 的安装与配置过程。

虽然 Gradle 在设计的时候基本沿用了 Maven 的这套依赖管理体系，但是在引用依赖时还是进行了一些改进。

（1）引用依赖的格式非常简洁，其内容只有依赖范围、组织名、包名、版本 4 个信息。

```
dependencies {
 compile 'org.hibernate:hibernate-core:3.6.7.Final'
 testCompile 'junit:junit:4.+'   //“+”表示支持动态版本管理
}
```

（2）Maven 和 Gradle 的依赖项的范围有所不同。在 Maven 中，一个依赖项有 6 种范围，分别是 complie（默认）、provided、runtime、test、system、import。而 Gradle 将其简化为 4 种，即 compile、runtime、testCompile、testRuntime。在后面的章节中我们会细述其使用场景。

（3）Gradle 支持动态的版本依赖。通过在版本号后面使用“+”的方式可以实现动态的版本管理。

（4）在解决依赖冲突方面，Gradle 的实现机制更加明确。在使用 Maven 和 Gradle 进行依赖管理时采用的都是传递性依赖，当多个依赖项指向同一个依赖项的不同版本时就会引起依赖冲突。Maven 处理这种依赖关系的能力十分有限，而 Gradle 在解决依赖冲突方面能力较强。

Gradle 和 Maven 都使用一致的项目结构。在 Ant 时代，大家创建 Java 项目目录时比较随意，然后通过 Ant 配置指定哪些项目属于 source，哪些项目属于 test source 等。而 Maven 设计之初的理念就是 Conversion Over Configuration（约定大于配置），其制定了一套项目目录结构作为标准的 Java 项目结构。一个典型的 Maven 项目结构如图 2-1 所示。

Gradle 也沿用了这一标准的目录结构，当然也可以在 Project Structure 下定义目录结构的属性。如果 Gradle 项目采用了标准的 Maven 项目结构，那么在 Gradle 中无须进行多余的配置，只需在文件中包含 apply plugin:'java'，系统就会自动识别 source、resource、test source、test resource 等相应资源。不过 Gradle 作为 JVM 上的构建工具，同时也支持 Groovy、Scala 等源代码的构建，甚至支持 Java、Groovy、Scala 语言的混合构建。虽然 Maven 通过一些插件（如 scala-plugin）也能达到相同目的，但在配置方面 Gradle 显然要更优雅简洁一些。

```
src
 +- main
     +- java
         +- com
             +- example
                +- myproject
                    +- Application.java
                    |
                    +- domain
                    |   +- Customer.java
                    |   +- CustomerRepository.java
                    |
                    +- service
                    |   +- CustomerService.java
                    |
                    +- web
                        +- CustomerController.java
     +- resources
 +- test
     +- java
     +- resources
```

图 2-1　Maven 项目结构

Maven 和 Gradle 设计时都采用了插件机制，但显然 Gradle 更胜一筹，其主要原因是 Maven 基于 XML 进行配置，其配置语法受限于 XML，即使实现很小的功能也需要设计一个插件，并建立该插件与 XML 配置的关联。例如，想在 Maven 中执行一条 shell 命令，其配置如下：

```
<plugin>
 <groupId>org.codehaus.mojo</groupId>
 <artifactId>exec-maven-plugin</artifactId>
 <version>1.2</version>
 <executions>
 <execution>
 <id>drop DB => db_name</id>
 <phase>pre-integration-test</phase>
 <goals>
 <goal>exec</goal>
 </goals>
 <configuration>
 <executable>curl</executable>
 <arguments>
 <argument>-s</argument>
 <argument>-S</argument>
 <argument>-X</argument>
 <argument>DELETE</argument>
 <argument>http://${db.server}:${db.port}/db_name</argument>
 </arguments>
 </configuration>
 </execution>
 </executions>
</plugin>
```

而在 Gradle 中这一切将变得非常简单，如下：

```
task dropDB(type: Exec) {
 commandLine ‘curl’,’-s’,’s’,’-x’,’DELETE’,"http://${db.server}:{db.port}/db_name"
}
```

本书第 5 章将教大家如何在 build.gradle 文字中使用 Gradle 的任务调用 MyBatis Generator 工具自动生成持久层。

在创建自定义插件方面，Maven 和 Gradle 的机制差不多，都是继承自插件基类，然后实现自定义插件的功能，这里不再展开说明。在第 5 章中，我们将细述如何配置插件和任务。

从以上几个方面可以看出Maven和Gradle的主要差异。Maven的设计核心Convention Over Configuration 被 Gradle 发扬光大，而 Gradle 的配置（代码）又超越了 Maven。在 Gradle 中不仅任何配置都可以作为代码被执行，而且还可以随时使用已有的 Ant 脚本、Java 类库、Groovy 类库来辅助完成程序的编写。

这种采用本身语言实现的 DSL 对本身语言项目进行构建管理的例子比比皆是，如 Rake 和 Ruby、Grunt 和 JavaScript、Sbt 和 Ruby 等。而 Gradle 之所以使用 Groovy 语言实现，是因为 Groovy 语言比 Java 语言更具表现力，其语法特性更为丰富，又兼具函数式的特点。这几年兴起的语言（如 Scala、Go、Swift）都属于强类型的语言，兼具面向对象和函数式的特点。

2.2 Gradle 的安装与配置

2.2.1 Gradle 的安装

访问 Gradle 的官网，下载 Gradle 3.4.1，解压后，在环境变量中添加系统变量 gradle_home，设置 Gradle 的解压路径为“E:\课程\软件\gradle-3.4.1-bin\gradle-3.4.1”，修改环境变量 Path 的值，具体如图 2-2 所示。

图 2-2　添加系统变量 gradle_home

下面在 IDEA 中配置 Gradle，选择菜单栏中的 File→Settings→Build,Execution,Deployment→Gradle，打开如图 2-3 所示界面。

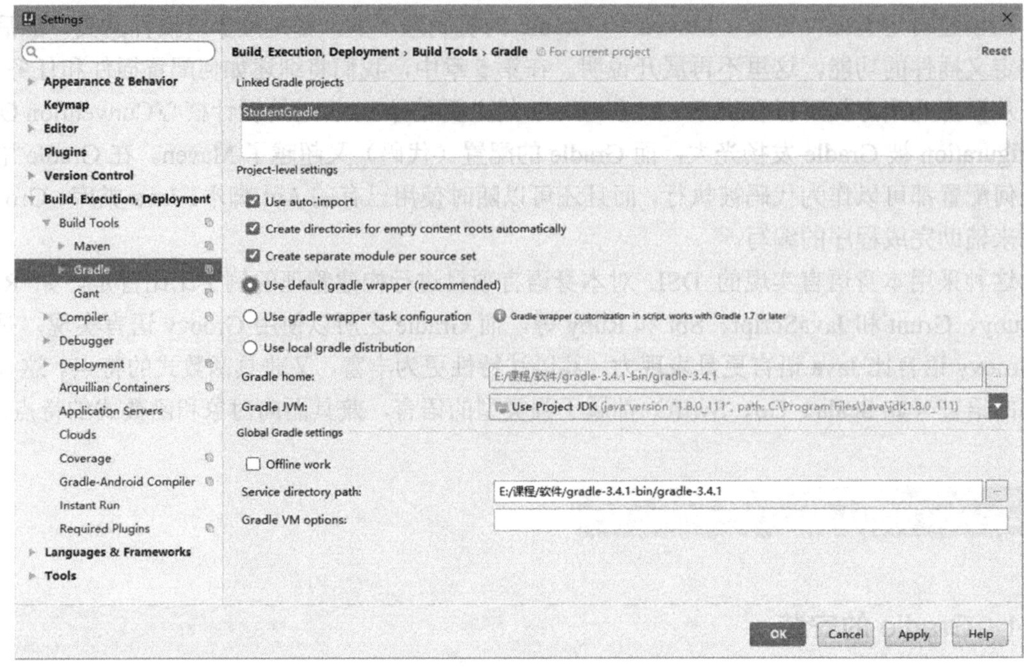

图 2-3　IDEA 平台下的 Gradle 配置界面

如果使用已安装好的Gradle软件，则选择Use local gradle distribution单选按钮，并在Gradle home 输入框中设置 Gradle 路径为“E:/课程/软件/gradle-3.4.1-bin/gradle-3.4.1”。如果选择 Use default gradle wrapper（recommended）单选按钮，则表示使用 IDEA 内置的 Gradle 插件。

2.2.2　Gradle 项目的目录结构

一个基于 Gradle 构建的项目目录结构如图 2-4 所示（Gradle 的配置文件 build.gradle 在图中已选中）。

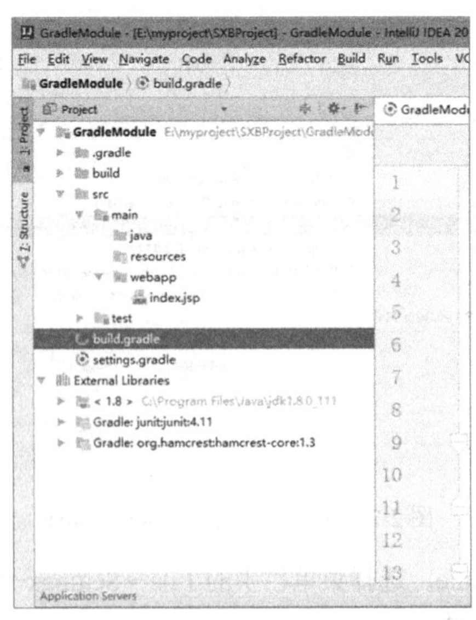

图 2-4　基于 Gradle 构建的项目目录结构

我们主要工作的文件夹是 src（source，源码），其下有 main 文件夹和 test 文件夹。

main 文件夹中的 java 文件夹用于存放 Java 源文件，如 bean、controller 等；resources 文件夹用于存放各种资源文件；webapp 文件夹用于存放 JSP、HTML、CSS、图片等网页文件。

test 文件夹用于存放自动生成的测试文件，其中 settings.gradle 文件在 Gradle 中用于初始化和配置项目树，存放在项目根目录下。Gradle 中多个项目以项目树形式表示，相当于 IDEA 中的 Project 和 Module 的概念。Project 为根项目，根项目下有很多子项目，即有很多 Module。Module 要被 Gradle 识别，需要在 settings.gradle 文件中配置声明，这样在项目构建时才会包含在其中。而 artifactId 在 Gradle 中是项目的名称，其值保存在 settings.gradle 文件的 rootProject.name 中。图 2-5 中的代码定义了一个名为 TestModule 的 Module。

Gradle 为每个定义的 Module 都指定了相应的文件目录，但在一般情况下，我们只做 Module 的声明而不指定它的目录位置。其原因是当不指定相应目录时，会默认目录位置和 settings.gradle 文件同级，寻找当前目录下跟指定 Module 同名的目录作为其目录，但是当找不到这个同名目录时就会报错。

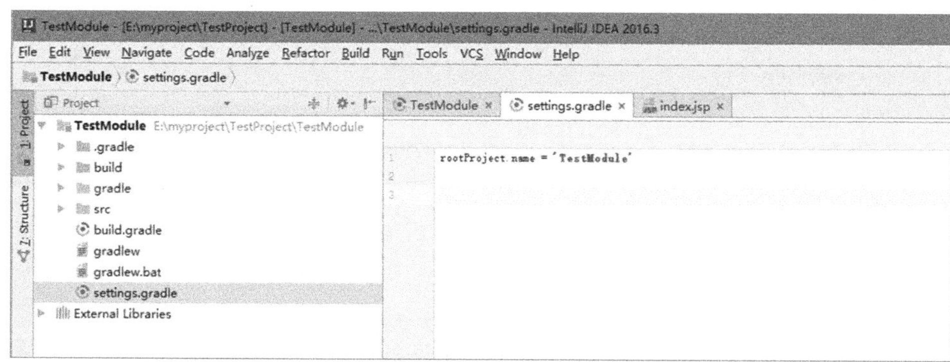

图 2-5　settings.gradle 文件的内容

2.2.3　build.gradle 文件

build.gradle 文件是 Gradle 模块中非常重要的配置文件，仓库和依赖都在该文件中定义。我们以项目中一个具体的 build.gradle 文件为例进行讲解，每一行代码有相应的注释，具体内容如下：

```
group 'com.alibaba'                          //项目组织名称
version '1.0-SNAPSHOT'                       //项目版本
apply plugin: 'java'                         //指定项目为 Java 项目，项目编译时生成项目的 jar 包
// 指定 Web 项目，项目编译（在项目提示符下执行：Gradle build）时生成项目的 war 包
apply plugin: 'war'
sourceCompatibility = 1.8                    //指定编译.java 文件时使用的 JDK 版本
ext {
    Spring_version = "4.3.14.RELEASE"  //定义常量 Spring_version
}

repositories {
    mavenCentral()                           //从 Maven 中央仓库中下载包
```

```
    maven {
        //从国内镜像阿里巴巴仓库下载 Maven，提高速度
        url 'http://maven.aliyun.com/nexus/content/groups/public/'
    }
}

dependencies {                                                    //项目依赖定义
    testCompile group: 'junit', name: 'junit', version: '4.11'//测试编译时使用 junit 包
    //编译时使用的 spring-web 包
    compile "org.springframework:spring-web:$Spring_version"
    ...
    }
```

在 IDEA 平台下打开 build.gradle 文件，详情如图 2-6 所示。

```
group 'person.xjl'
version '1.0-SNAPSHOT'

apply plugin: 'java'
apply plugin: 'war'

sourceCompatibility = 1.8

repositories {
  maven { url "http://maven.aliyun.com/nexus/content/groups/public/" }
   mavenCentral()
}

dependencies {
   testCompile group: 'junit', name: 'junit', version: '4.11'
  }
```

图 2-6　build.gradle 文件内容

在 build.gradle 文件中，dependencies（依赖）有好几种类型，分别是 testCompile、compile、providedCompile 及 runtime。

如果 jar 包或库不仅在项目编译的时候被需要，而且在运行的时候也被需要，那么就使用 providedCompile。例如：

```
providedCompile 'org.springframework:spring-webmvc:4.3.14.RELEASE'
```

如果 jar 包或库仅在项目编译的时候被需要，而在运行时不被需要，那么就使用 compile。例如：

```
compile group: 'javax.servlet.jsp.jstl', name: 'jstl', version: '1.1'
```

如果 jar 包或库仅在运行的时候被需要，而在编译时不被需要，那么就使用 runtime。例如：

```
runtime group: 'mysql', name: 'mysql-connector-java', version: '5.1.42'
```

testCompile 顾名思义就是测试时需要的依赖，如 junit。

以上这些依赖，需要正确配置 apply plugin: 'java' 或者 apply plugin: 'war'，否则会遇到依赖包无法导入，以及 runtime 和 providedCompile 无法使用的情况。

注 意 如果 Maven 中央仓库下载速度慢，则可以使用国内镜像阿里巴巴仓库。具体使用方法是，将下面这段代码复制到 repositories 标签内。

```
maven {
        //从国内镜像阿里巴巴仓库下载 Maven，提高速度
        url 'http://maven.aliyun.com/nexus/content/groups/public/'
}
```

2.3 Gradle 项目工作任务

2.3.1 任务一：创建 Gradle 构建的项目

1. 创建项目

步骤 1．选择菜单栏中的 File→New→Project→Empty Project，弹出 New Project 对话框，如图 2-7 所示。

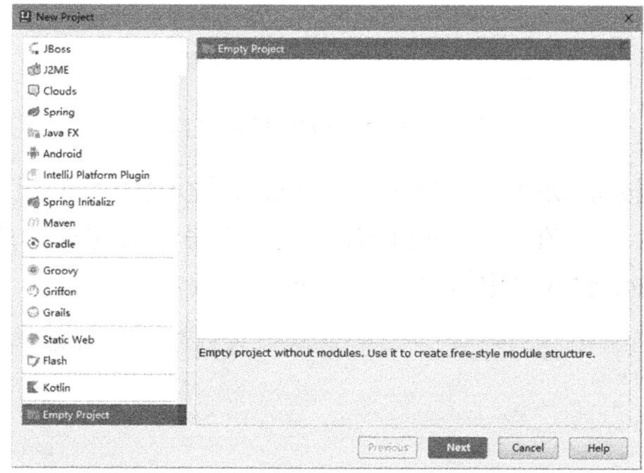

图 2-7 创建项目

步骤 2．单击 Next 按钮，在 Project name 输入框中填写项目名，在 Project location 输入框中设置项目存放路径，如图 2-8 所示。

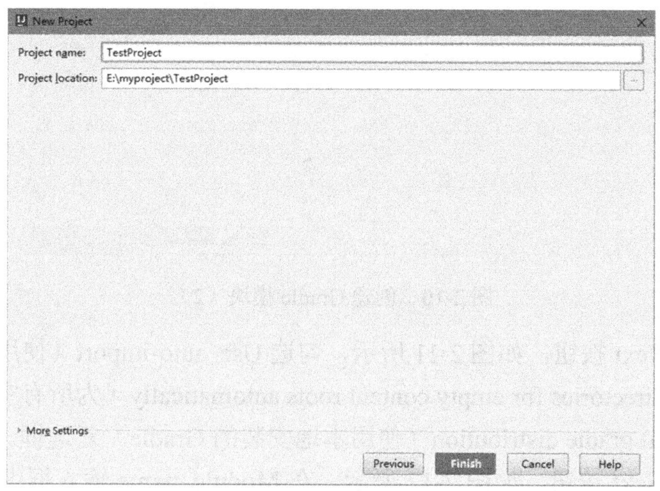

图 2-8 设置项目名和存放路径

2．基于 Gradle 构建 Module

步骤 1．选择菜单栏中的 File→New→Module→Gradle，弹出 New Module 对话框，如图 2-9 所示。在弹出的对话框中选择 Gradle 选项后，勾选 Java 复选框和 Web 复选框（为该 Module 增加本地的 jdk 包和 Web 包，并自动创建 Web 项目的目录结构）。

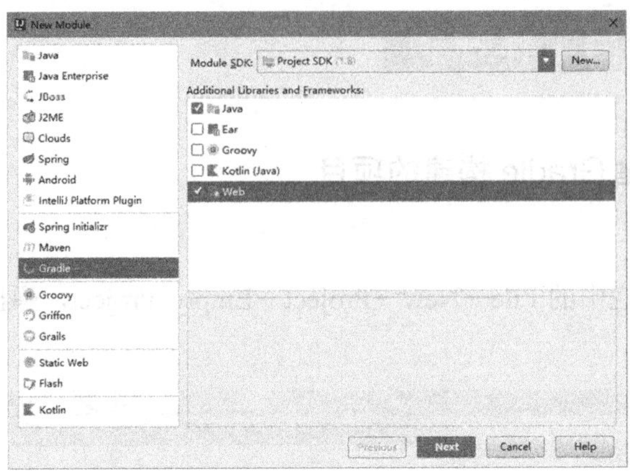

图 2-9　创建 Gradle 模块（1）

步骤 2．单击 Next 按钮，如图 2-10 所示，在 GroupId 输入框中填写组织名称，如 org.apache、com.alibaba、person.xjl 等；在 ArtifactId 输入框中填写最后完成的成品名称，如 Lego、StudentGradle；在 Version 输入框中填写当前 Module 的版本号。

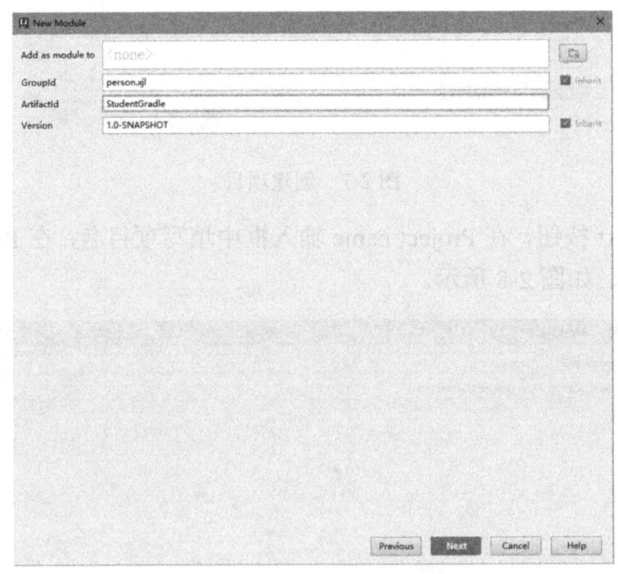

图 2-10　创建 Gradle 模块（2）

步骤 3．单击 Next 按钮，如图 2-11 所示，勾选 Use auto-import（使用自动导入功能）复选框，勾选 Create directories for empty content roots automatically（为所有空内容创建目录）复选框，勾选 Use local gradle distribution（使用本地安装的 Gradle）复选框。

步骤 4．单击 Next 按钮，如图 2-12 所示，在 Module name 输入框中填写模块名称，在 Content root 输入框中填写模块存放路径。

步骤5. 单击 Finish 按钮，当 IDEA 界面中状态条不闪烁时，Gradle 项目即构建完成。

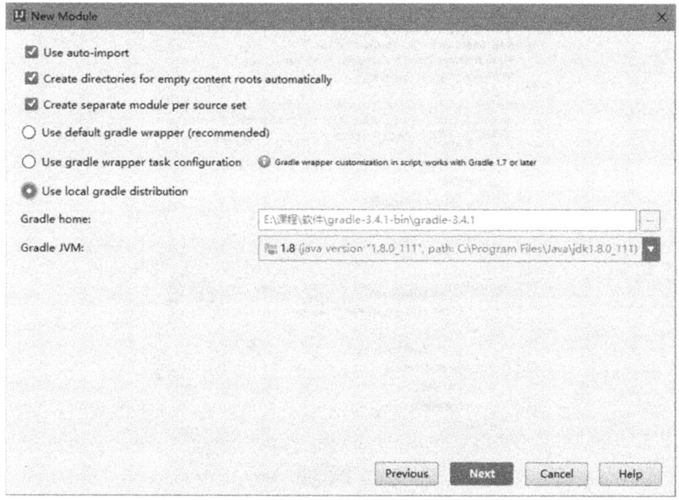

图 2-11 创建 Gradle 模块（3）

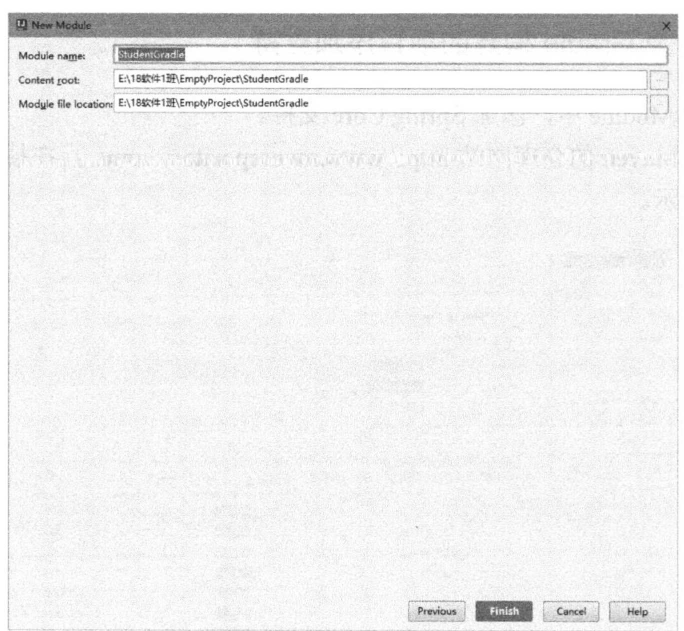

图 2-12 创建 Gradle 模块（4）

2.3.2 任务二：导入 Gradle 构建的项目

导出：直接在 Module 所在文件夹路径下打包。

导入：解压打包项目文件后，在 IDEA 主界面中选择菜单栏中的 File→Project Structure→Modules→“+”→build.gradle，在弹出的对话框中设置导入文件的路径，如图 2-13 所示。

注 意 在导入时一定要选择 build.gradle 文件。如果是用 Maven 构建的 Web 项目，则选择 pom.xml 文件。

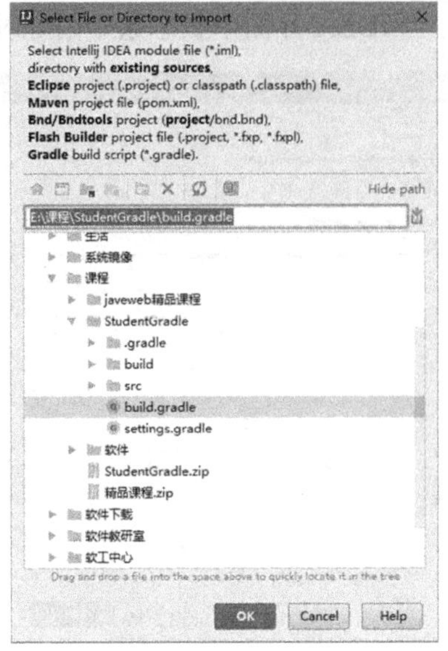

图 2-13　导入文件

2.3.3　任务三：为 Gradle 构建的项目添加支持

在上面创建的 Module 中，添加 Spring Core 支持。

步骤 1．打开 Maven 的仓库网站 http://www.mvnrepository.com/ ，在搜索框中搜索 spring core，如图 2-14 所示。

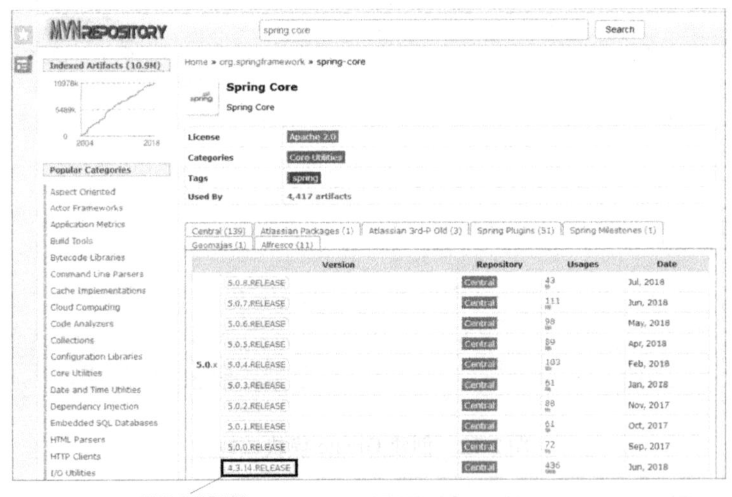

图 2-14　搜索 spring core

步骤 2．找到 Usages（使用率）相对较高的版本，即 4.3.14.RELEASE，单击该按钮。

步骤 3．选择 Gradle 选项卡，如图 2-15 所示，将里面的内容粘贴到 Gradle 的配置文件 build.gradle 的 dependencies 选项卡里。

拓展练习：添加 jstl 1.2 版本的依赖。

图 2-15　Gradle 选项卡

2.4　Gradle 构建项目的管理

在 IDEA 主界面中，选择菜单栏中的 File→Project Structure 或者单击主界面右上角的按钮，都将弹出如图 2-16 所示的 Project Structure 对话框，在其中对 Project 和 Module 的结构和目录进行管理。

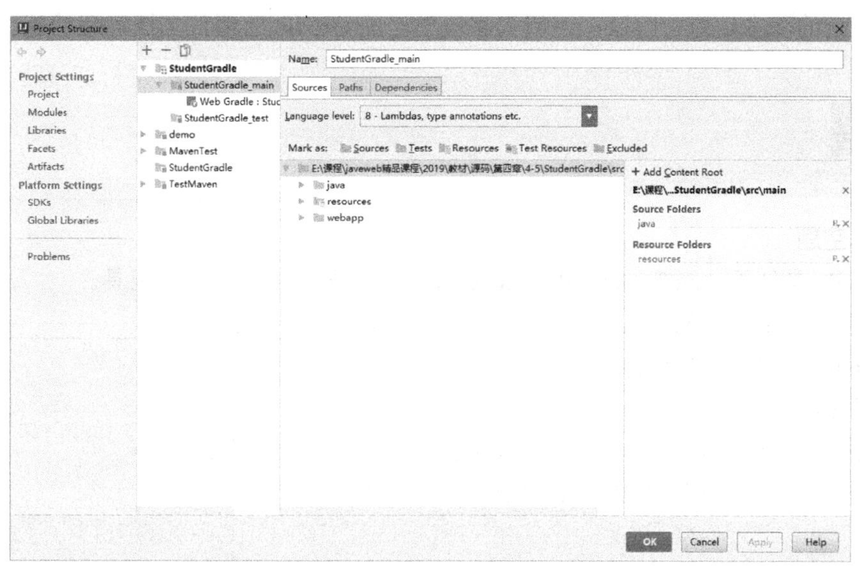

图 2-16　对项目和模块的管理

对基于 Gradle 构建的 Module 的管理在右侧边栏进行，如图 2-17 所示。如果 IDEA 主界面没有右侧边栏，可通过单击 IDEA 主界面左下角的显示器图标使其显示，详情请参考图 1-2。

项目下所有基于Gradle构建的Module都会在Gradle projects 窗口中显示，可以通过图 2-17 中左上角的按钮刷新 Module 的依赖。

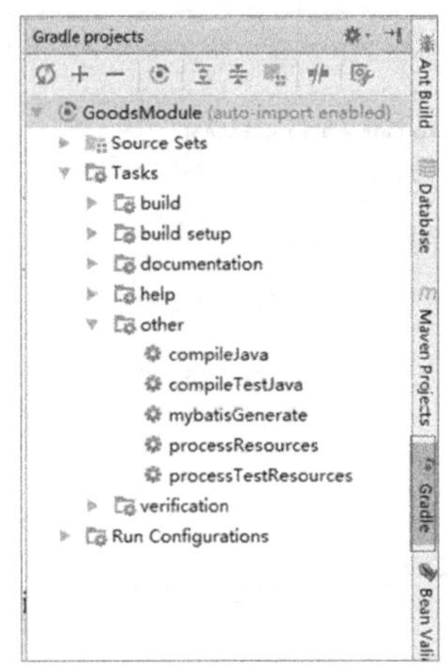

图 2-17　基于 Gradle 构建的 Module 的管理

小　结

Gradle 作为新兴的项目自动化构建工具，具有简便、实用的特点。它沿用 Maven 的中央仓库，用其编写的代码相对简单易读，故它将逐步成为业界的主流构建工具。我们需要熟练掌握 build.gradle 文件的编写方法，主要涉及仓库位置确定、依赖管理、插件配置等内容，以及能熟练导入和导出基于 Gradle 构建的 Module。本书所有案例都以本章案例为基础，最后实现学生管理系统的基本功能。所以，希望大家自学一些 Gradle 的项目文档管理功能。

习　题

一、填空题

1．Gradle 是基于 Apache Ant 和 Apache Maven 概念的__________工具。

2．目前市面上流行的三大项目自动化工具分别是_______、________、________。

3．Gradle 配置文件的名称是___________。

二、简答题

1．简述 Gradle 构建工具与 Maven 构建工具的区别与相似之处。

2．简述 Gradle 如何管理项目所需要的第三方包。

综合实训

实训 1．创建一个 Gradle 的模块，并配置 Spring MVC 框架，在 Tomcat 服务器上运行。

实训 2．为实训 1 添加 jstl 依赖支持。

实训 3．实现模块的导入，并部署访问，同时将过程截图。

第3章

项目前端框架集成

本章学习目标

- 了解前端框架 Bootstrap 和 EasyUI
- 熟悉将前端框架集成到 Gradle 项目的过程
- 掌握前端框架组件在项目中的使用方法

本章简单介绍 Bootstrap 前端框架的特性与使用方法，着重讲解如何将其集成到第 2 章创建的基于 Gradle 的项目中，并使用 Bootstrap 和 EasyUI 框架的复杂组件和布局构建网页。

3.1 Bootstrap 简介

Bootstrap 是 Twitter 推出的最受欢迎的 HTML、CSS 和 JS 框架，用于开发响应式布局、移动设备优先的 Web 项目。

Bootstrap 让前端开发更快速、简单，所有开发者都能快速上手，所有设备都可以适配，所有项目都适用。

Bootstrap 是完全开源的，于 2011 年 8 月发布在 GitHub 平台上。它的代码托管、开发、维护都依赖 GitHub 平台。

3.2 Bootstrap 的集成与使用

为了能在第 2 章创建的基于 Gradle 的 Module 中使用 Bootstrap 框架的表单或组件，需要下载 Bootstrap 框架并将其集成到已创建的模块中。下面就讲解其详细过程和步骤。

3.2.1 Bootstrap 的下载与集成

访问 Bootstrap 官网，如图 3-1 所示，单击“下载 Bootstrap”超链接，下载 3.3.7 版本的 Bootstrap 压缩包。

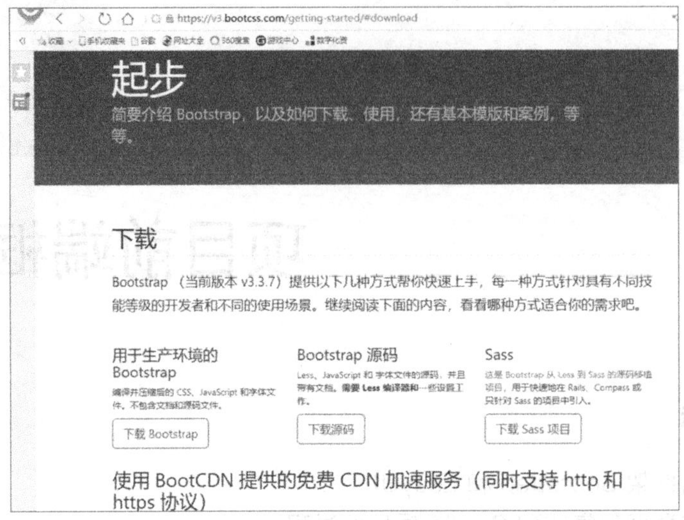

图 3-1　Bootstrap 官网下载界面

将 Bootstrap 压缩包解压后，将得到 3 个子文件夹：css、fonts 和 js。这 3 个子文件夹分别存放了 Bootstrap 所需要的布局、字体和 JS 文件。图 3-2 展示的就是解压后的 Bootstrap 文件结构，里面有编译好的 CSS 和 JS（bootstrap.*）文件，还有经过压缩的 CSS 和 JS（bootstrap.min.*）文件，以及可以在某些浏览器的开发工具中使用的 CSS 源码映射表（bootstrap.*.map），还包含了来自 Glyphicons 的 Bootstrap 主题中所使用的图标字体。

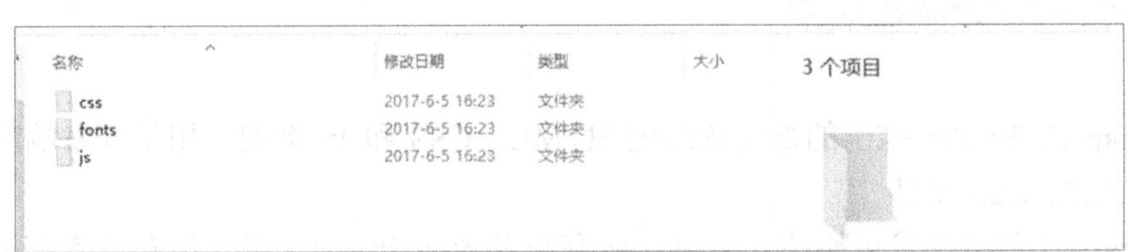

图 3-2　解压后的 Bootstrap 文件结构

将解压后的文件和文件夹复制到 TestModule 的 webapp 文件夹下，如图 3-3 所示。另外，还需要下载一个 jquery- 3.3.1.min.js 文件放置于 js 文件夹下。

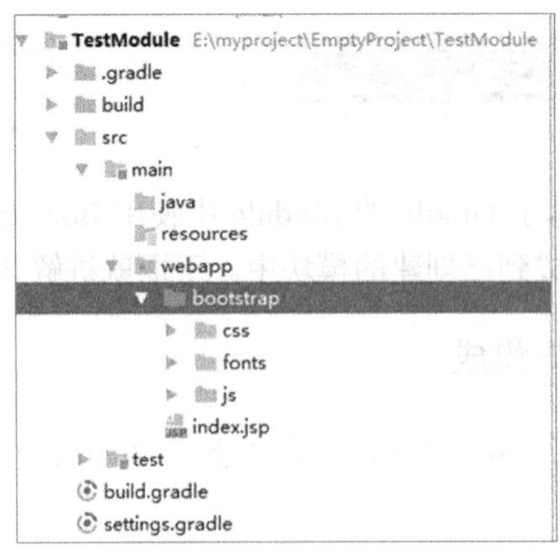

图 3-3　复制 bootstrap 文件夹到 webapp 文件夹

3.2.2 Bootstrap 框架组件的使用

在 JSP 文件中使用 Bootstrap 组件前，必须先引用链接.css，并声明使用的脚本文件。

步骤 1．声明引用。

```
 <link href="bootstrap/css/bootstrap.min.css" rel="stylesheet">
<!-- jQuery 文件。务必在 bootstrap.min.js 之前引入 ,到 jquery 官网上下载最新的 juery 版本-->
<script src="bootstrap/js/jquery-3.1.1.min.js"></script>
<!-- 最新的 Bootstrap 核心 JavaScript 文件 -->
<script src="bootstrap/js/bootstrap.min.js"></script>
```

步骤 2．使用 Bootstrap 组件——条纹状表格，如图 3-4 所示。

图 3-4 “条纹状表格”界面

将下面这段代码放到 JSP 文件的<body>和</body>标签之间即可。

```
<table class="table table-striped">
 <caption>条纹表格布局</caption>
 <thead>
  <tr>
   <th>#</th>
   <th>First Name</th>
   <th>Last Name</th>
<th>Username</th>

  </tr>
 </thead>
 <tbody>
  <tr>
   <td>1</td>
   <td>Mark</td>
   <td>Otto</td>
   <td>@mdo</td>
  </tr>
 ... </tbody>
</table>
```

3.3 Bootstrap 框架的使用

3.3.1 任务一：完成登录界面的设计

登录界面效果如图 3-5 所示。

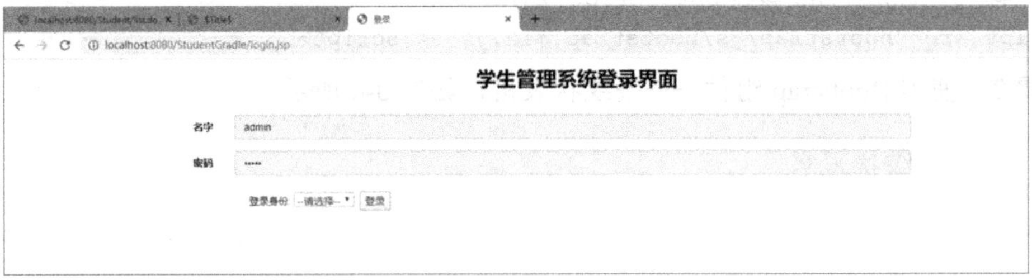

学生管理系统登录界面

图 3-5 登录界面效果

在 webapp 文件夹下新建 login.jsp 文件，源文件内容如下：

```
<%@ page language="java" import="java.util.*" pageEncoding="UTF-8"%>
<%
    String path = request.getContextPath();
    String basePath = request.getScheme() + "://"
            + request.getServerName() + ":" + request.getServerPort()
            + path + "/";
%>
<!DOCTYPE HTML PUBLIC "-//W3C//DTD HTML 4.01 Transitional//EN">
<html>
<head>
    <link href="bootstrap/css/bootstrap.min.css" rel="stylesheet">
    <!-- jQuery 文件。务必在 bootstrap.min.js 之前引入 -->
    <script src="bootstrap/js/jquery-3.3.1.min.js"></script>
    <!-- 最新的 Bootstrap 核心 JavaScript 文件 -->
    <script src="bootstrap/js/bootstrap.min.js"></script>
    <base href="<%=basePath%>">
    <title>登录</title>
    <meta http-equiv="pragma" content="no-cache">
    <meta http-equiv="cache-control" content="no-cache">
    <meta http-equiv="expires" content="0">
    <meta http-equiv="keywords" content="keyword1,keyword2,keyword3">
    <meta http-equiv="description" content="This is my page">
    <!--<link rel="stylesheet" type="text/css" href="styles.css">-->
</head>
<script type="text/javascript">
    function check_login() {
        var username = document.getElementById("username");
        var password = document.getElementById("password");
        var type = document.getElementById("type");
        if(type.value == "-1"){
```

```
            window.alert("请选择登录身份，不得为空！");
            return false;
        }else{
            if (username.value == "" || password.value == "") {
                window.alert("登录 ID、登录密码都不能为空！");
                return false;
            }
            return true;
        }
    }
</script>
<body >
<form class="form-horizontal" role="form" action="check" method="post">
    <div class="form-group" align="center">
        <div class="form-group" align="center">
            <label  class="col-sm-6 control-label" style="font-size: 28px; color:black;
font-family: 华文楷体;" align="center">学生管理系统登录界面</label>
        </div>
    </div>
    <div class="form-group">
        <label for="username" class="col-sm-2 control-label" style=" color:black;">
名字</label>
        <div class="col-sm-6">
            <input type="text" class="form-control" name="username" id="username"
placeholder="请输入名字">
        </div>
    </div>
    <div class="form-group">
        <label for="password" class="col-sm-2 control-label" style=" color:black;">
密码</label>
        <div class="col-sm-6">
            <input type="password" class="form-control" name="password" id="password"
placeholder="请输入密码">
        </div>
    </div>
    <div class="form-group">
        <div class="col-sm-offset-2 col-sm-10">
            <div class="checkbox">
                <td class="s2">
                    <label style=" color:black;">
                        登录身份：
                    </label>
                </td>
                <td>
                    <select id="type" name="type">
                        <option value="-1">
                            --请选择--
                        </option>
```

```
                    <option value="0">
                        学生
                    </option>
                    <option value="1">
                        教师
                    </option>
                    <option value="2">
                        管理员
                    </option>
                </select>
            </td>
            <td height="37" colspan="2" align="right">
                 <button type="submit" id="submit" name="submit" value="登
录">登录</button>
                 <button type="reset" id="reset" name="reset" style= "background-
color:transparent;border: 0;" ></button>
            </td>
        </div>
    </div>
  </div>
</form>
</body>
</html>
```

3.3.2 任务二：使用扩展日历时间组件 datetimepicker

步骤 1．访问 http://www.bootcss.com/p/bootstrap-datetimepicker/，单击“下载”按钮，下载 bootstrap-datetimepicker-master.zip 压缩包。

步骤 2．将其解压后，复制 css 文件夹和 js 文件夹到项目的 webapp/bootstrap 文件夹下，完成 datetimepicker 组件的添加。

步骤 3．在 example-dtp.jsp 页面中使用该组件，源码如下：

```
<!DOCTYPE html>
<html>
<head>
    <title></title>
    <link href="./bootstrap/css/bootstrap.min.css" rel="stylesheet" media= "screen">
    <link href="../bootstrap/css/bootstrap-datetimepicker.css" rel="stylesheet" media=
"screen">
</head>

<body>
<div class="container">
    <form action="" class="form-horizontal"  role="form">
        <fieldset>
            <legend>Test</legend>
```

```
            <div class="form-group">
                <label for="dtp_input1" class="col-md-2 control-label">DateTime Picking
</label>
                <div class="input-group date form_datetime col-md-5" data-date= "1979-
09-16T05:25:07Z" data-date-format="dd MM yyyy - HH:ii p" data-link-field= "dtp_input1">
                    <input class="form-control" size="16" type="text" value=""
 readonly>
                    <span class="input-group-addon"><span class="glyphicon glyphicon-
remove"></span></span>
                    <span class="input-group-addon"><span class="glyphicon glyphicon-
th"></span></span>
                </div>
                <input type="hidden" id="dtp_input1" value="" /><br/>
            </div>
            <div class="form-group">
                <label for="dtp_input2" class="col-md-2 control-label">Date Picking
</label>
                <div class="input-group date form_date col-md-5" data-date="" data-
date-format="dd MM yyyy" data-link-field="dtp_input2" data-link-format= "yyyy-mm-dd">
                    <input class="form-control" size="16" type="text" value=""
readonly>
                    <span class="input-group-addon"><span class="glyphicon glyphicon-
remove"></span></span>
                    <span class="input-group-addon"><span class="glyphicon glyphicon-
calendar"></span></span>
                </div>
                <input type="hidden" id="dtp_input2" value="" /><br/>
            </div>
            <div class="form-group">
                <label for="dtp_input3" class="col-md-2 control-label">Time Picking
</label>
                <div class="input-group date form_time col-md-5" data-date="" data-
date-format="hh:ii" data-link-field="dtp_input3" data-link-format="hh:ii">
                    <input class="form-control" size="16" type="text" value=""
readonly>
                    <span class="input-group-addon"><span class="glyphicon glyphicon-
remove"></span></span>
                    <span class="input-group-addon"><span class="glyphicon glyphicon-
time"></span></span>
                </div>
                <input type="hidden" id="dtp_input3" value="" /><br/>
            </div>
        </fieldset>
    </form>
</div>
```

```
    <script  type="text/javascript"  src="bootstrap/js/jquery-3.3.1.min.js"  charset=
"UTF-8"></script>
    <script type="text/javascript" src="bootstrap/js/bootstrap.min.js"></script>
    <script   type="text/javascript"   src="bootstrap/js/bootstrap-datetimepicker.js"
charset= "UTF-8"></script>
    <script type="text/javascript" src="bootstrap/js/locales/bootstrap-datetimepicker.
zh-CN.js" charset="UTF-8"></script>
    <script type="text/javascript">
        $('.form_datetime').datetimepicker({
            language:  'zh-CN',
            weekStart: 1,
            todayBtn:  1,
            autoclose: 1,
            todayHighlight: 1,
            startView: 2,
            forceParse: 0,
            showMeridian: 1
        });
        $('.form_date').datetimepicker({
            language:  'zh-CN',
            weekStart: 1,
            todayBtn:  1,
            autoclose: 1,
            todayHighlight: 1,
            startView: 2,
            minView: 2,
            forceParse: 0
        });
        $('.form_time').datetimepicker({
            language:  'zh-CN',
            weekStart: 1,
            todayBtn:  1,
            autoclose: 1,
            todayHighlight: 1,
            startView: 1,
            minView: 0,
            maxView: 1,
            forceParse: 0
        });
    </script>

    </body>
    </html>
```

步骤 4．扩展日历时间组件运行效果如图 3-6 所示。

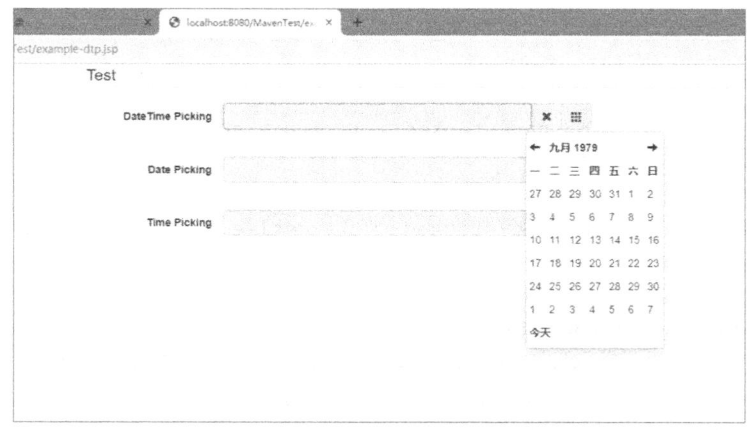

图 3-6　扩展日历时间组件运行效果

拓展练习：练习使用 Bootstrap 的其他组件，如导航条、轮播等。

3.3.3　任务三：左侧树状导航条的实现

左侧树状导航条的实现代码如下：

```
<%@ page contentType="text/html; charset=utf-8" language="java" %>
<%@ taglib prefix="c" uri="http://java.sun.com/jsp/jstl/core"%>
<html>
<head>
    <meta http-equiv="Content-Type" content="text/html; charset=utf-8" />
    <title></title>
    <link    href="http://cdn.bootcss.com/bootstrap/3.2.0/css/bootstrap.min.css"
rel="stylesheet">
    <style>
        #main-nav {
            margin-left: 1px;
        }
        #main-nav.nav-tabs.nav-stacked > li > a {
            padding: 10px 8px;
            font-size: 12px;
            font-weight: 600;
            color: #4A515B;
            background: #E9E9E9;
            background: -moz-linear-gradient(top, #FAFAFA 0%, #E9E9E9 100%);
            background:  -webkit-gradient(linear,  left  top,  left  bottom,  color-
stop(0%,#FAFAFA), color-stop(100%,#E9E9E9));
            background: -webkit-linear-gradient(top, #FAFAFA 0%,#E9E9E9 100%);
            background: -o-linear-gradient(top, #FAFAFA 0%,#E9E9E9 100%);
            background: -ms-linear-gradient(top, #FAFAFA 0%,#E9E9E9 100%);
            background: linear-gradient(top, #FAFAFA 0%,#E9E9E9 100%);
            filter:     progid:DXImageTransform.Microsoft.gradient(startColorstr=
'#FAFAFA', endColorstr='#E9E9E9');
            -ms-filter: "progid:DXImageTransform.Microsoft.gradient(startColorstr=
'#FAFAFA', endColorstr='#E9E9E9')";
```

```
            border: 1px solid #D5D5D5;
            border-radius: 4px;
        }
        #main-nav.nav-tabs.nav-stacked > li > a > span {
            color: #4A515B;
        }
        #main-nav.nav-tabs.nav-stacked > li.active > a, #main-nav.nav-tabs.nav-
stacked > li > a:hover {
            color: #FFF;
            background: #3C4049;
            background: -moz-linear-gradient(top, #4A515B 0%, #3C4049 100%);
            background: -webkit-gradient(linear, left top, left bottom, color-stop
(0%,#4A515B), color-stop(100%,#3C4049));
            background: -webkit-linear-gradient(top, #4A515B 0%,#3C4049 100%);
            background: -o-linear-gradient(top, #4A515B 0%,#3C4049 100%);
            background: -ms-linear-gradient(top, #4A515B 0%,#3C4049 100%);
            background: linear-gradient(top, #4A515B 0%,#3C4049 100%);
            filter:     progid:DXImageTransform.Microsoft.gradient(startColorstr=
'#4A515B', endColorstr='#3C4049');
            -ms-filter: "progid:DXImageTransform.Microsoft.gradient(startColorstr=
'#4A515B', endColorstr='#3C4049')";
            border-color: #2B2E33;
        }
        #main-nav.nav-tabs.nav-stacked > li.active > a, #main-nav.nav-tabs.nav-
stacked > li > a:hover > span {
            color: #FFF;
        }
        #main-nav.nav-tabs.nav-stacked > li {
            margin-bottom: 4px;
        }
        /*定义二级菜单样式*/
        .secondmenu a {
            font-size: 10px;
            color: #4A515B;
            text-align: center;
        }
        .navbar-static-top {
            background-color: #212121;
            margin-bottom: 5px;
        }
        .navbar-brand {
            background: url('') no-repeat 10px 8px;
            display: inline-block;
            vertical-align: middle;
            padding-left: 50px;
            color: #fff;
        }
    </style>
</head>
<body style="background-color: #dee2e6">
```

```
<div class="container-fluid">
    <div class="row">
        <div class="col-md-2">
            <ul id="main-nav" class="nav nav-tabs nav-stacked" style="">
                <li class="active">
                    <a href="/welcome" target="main">
                        <i class="glyphicon glyphicon-th-large"></i>
                        首页
                    </a>
                </li>
                <li>
                    <a href="#systemSetting" class="nav-header collapsed" data-toggle
="collapse">
                        <i class="glyphicon glyphicon-cog"></i>
                        学生管理
                        <span class="pull-right glyphicon glyphicon-chevron-down">
</span>
                    </a>
                    <ul id="systemSetting" class="nav nav-list collapse secondmenu"
style="height: 0px;">
                        <li><a href="student/list" target="main"><i class="glyphicon
glyphicon-user"></i>学生信息</a></li>
                        <%--<li><a href="#"><i class="glyphicon glyphicon-th-list">
</i>菜单管理</a></li>--%>
                        <%--<li><a href="#"><i class="glyphicon glyphicon-asterisk">
</i>角色管理</a></li>--%>
                        <%--<li><a href="#"><i class="glyphicon glyphicon-edit"></i>
修改密码</a></li>--%>
                        <%--<li><a href="#"><i class="glyphicon glyphicon-eye-open">
</i>日志查看</a></li>--%>
                    </ul>
                </li>
                <li>
                    <a href="course/list" target="main">
                        <i class="glyphicon glyphicon-credit-card"></i>
                        课程管理
                    </a>
                </li>
                <li>
                    <a href="teacher/list" target="main">
                        <i class="glyphicon glyphicon-globe"></i>
                        教师管理
                        <span class="label label-warning pull-right"></span>
                    </a>
                </li>
                <li>
                    <a href="score/list" target="main">
                        <i class="glyphicon glyphicon-calendar"></i>
                        分数管理
                    </a>
                </li>
```

```
                    <li>
                        <a  href="#system"  class="nav-header  collapsed"  data-toggle=
"collapse">
                            <i class="glyphicon glyphicon-fire"></i>
                            管理员管理
                            <span  class="pull-right  glyphicon  glyphicon-chevron-down">
</span>
                        </a>
                        <ul   id="system"   class="nav   nav-list   collapse   secondmenu"
style="height: 0px;">
                            <li><a  href="admin/list"  target="main"><i  class="glyphicon
glyphicon-user"></i>管理员信息</a></li>
                            <li><a href="Admin!add.do" target="main"><i class="glyphicon
glyphicon-th-list"></i>添加管理员</a></li>
                            <li><a  href="admin/updatePassword"  target="main"><i  class=
"glyphicon glyphicon-edit"></i>修改密码</a></li>
                        </ul>
                    </li>
                </ul>
            </div>
            <div class="col-md-10">
            </div>
        </div>
    </div>
    <!-- jQuery 文件。务必在 bootstrap.min.js 之前引入 -->
    <script src="/bootstrap/js/jquery-3.3.1.min.js"></script>
    <!-- 最新的 Bootstrap 核心 JavaScript 文件 -->
    <script src="/bootstrap/js/bootstrap.min.js"></script>
    <script>
    </script>
    </body></html>
```

左侧树状导航条运行效果如图 3-7 所示。

图 3-7　左侧树状导航条运行效果

3.4　集成 EasyUI 前端框架

3.4.1　EasyUI 简介

EasyUI 是一种基于 jQuery 的用户界面插件集合。EasyUI 是一个完美支持 HTML 5 网页的前端框架。使用 EasyUI 框架可以创建具有现代感且交互能力强的动态网站。使用 EasyUI 不需要编写很多代码，只需要编写一些简单的 HTML 标记，就可以完成用户界面，因此可以缩短网页开发的时间。

3.4.2　EasyUI 的下载

（1）访问 EasyUI 官网，如图 3-8 所示，单击“下载”按钮，下载免费版 EasyUI 压缩包。

图 3-8　EasyUI 官网

（2）将 EasyUI 压缩包解压，解压后的文件夹和文件如图 3-9 所示。

demo	2016-8-25 9:02	文件夹	
demo-mobile	2015-11-22 17:26	文件夹	
locale	2018-2-13 22:34	文件夹	
plugins	2017-2-18 10:32	文件夹	
src	2018-3-23 11:01	文件夹	
themes	2018-1-24 23:29	文件夹	
changelog	2018-2-23 23:14	文本文档	35 KB
easyloader	2018-2-23 23:40	JavaScript 文件	7 KB
jquery.easyui.min	2018-2-23 23:40	JavaScript 文件	418 KB
jquery.easyui.mobile	2018-2-23 23:40	JavaScript 文件	5 KB
jquery.min	2015-6-18 10:18	JavaScript 文件	94 KB
license_freeware	2017-8-28 8:12	文本文档	2 KB
readme	2018-2-23 23:39	文本文档	1 KB

图 3-9　EasyUI 解压后的文件夹和文件

3.4.3 EasyUI 的集成与使用

步骤 1．复制 EasyUI 解压后的部分文件和文件夹到项目 StudentGradle 中，删掉 demo、demo-mobile、src 三个文件夹即可，如图 3-10 所示。

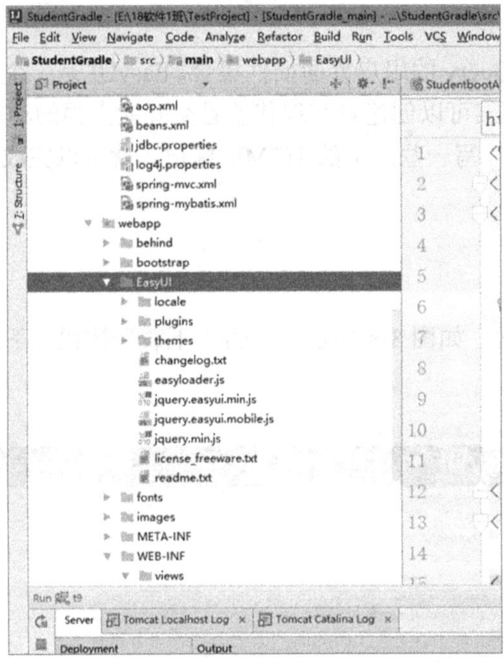

图 3-10 复制 EasyUI 的部分文件和文件夹到 StudentGradle

步骤 2．修改 resources/spring-mvc.xml 文件，增加静态资源映射的定义。

```
<mvc:resources mapping="EasyUI/**" location="EasyUI/" />
```

3.4.4 任务四：使用 EasyUI 组件导航树和对话框

步骤 1．在 JSP 页面中的<head>标签里添加以下代码：

```
<link rel="stylesheet" type="text/css" href="EasyUI/themes/default/easyui.css">
<link rel="stylesheet" type="text/css" href="EasyUI/themes/icon.css">
<script type="text/javascript" src="EasyUI/jquery.min.js"></script>
<script type="text/javascript" src="EasyUI/jquery.easyui.min.js"></script>
```

步骤 2．使用组件有如下两种方式。

（1）直接在 HTML 标签中声明组件，相应代码如下：

```
<div id="dd" class="EasyUI-dialog" title="My Dialog" style="width:400px;height:200px;" data-options="iconCls:'icon-save',resizable:true,modal:true"> Dialog Content.</div>
```

（2）编写 JavaScript 代码创建组件，相应代码如下：

```
<input id="cc" style="width:200px" />
$('#cc').combobox({
url: ...,
```

```
    required: true,
    valueField: 'id',
    textField: 'text'
    });
```

这里使用第一种方式，将一个对话框和一个导航树放在页面上，把下面这段代码复制到<body>标签中：

```
    <div id="dd" class="EasyUI-dialog" title="My Dialog" style="width:400px;height:
200px;" data-options="iconCls:'icon-save',resizable:true,modal:true"> Dialog Content.
</div>
    <ul class="easyui-tree">
        <li>
            <span>Folder</span>
            <ul>
                <li>
                    <span>Sub Folder 1</span>
                    <ul>
                        <li><span>File 11</span></li>
                        <li><span>File 12</span></li>
                        <li><span>File 13</span></li>
                    </ul>
                </li>
                <li><span>File 2</span></li>
                <li><span>File 3</span></li>
            </ul>
        </li>
        <li><span>File21</span></li>
    </ul>
```

步骤 3．部署运行，EasyUI 组件运行效果如图 3-11 所示。

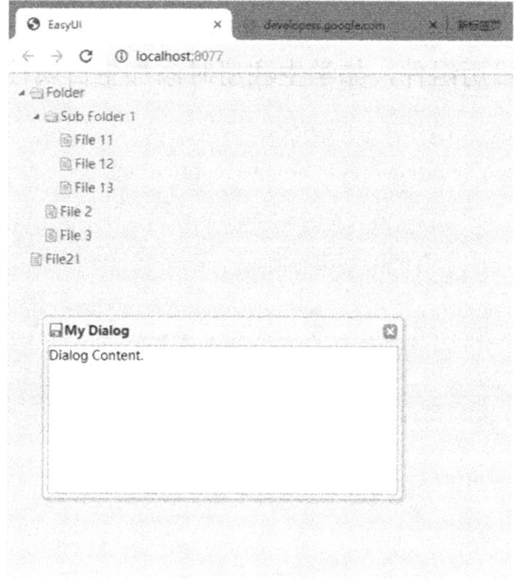

图 3-11　EasyUI 组件运行效果

拓展练习：练习使用 EasyUI 的其他组件，如表格、表单等。

小 结

使用前端框架可以极大地简化程序员的工作，快速地构造出大气、清晰的网站架构和后台管理系统。并且，前端框架的配置和集成都非常方便。本章的重点是学习将前端框架 Bootstrap 和 EasyUI 集成到 Gradle 构建的项目中，并能灵活使用一些复杂的组件。

习 题

一、填空题

1．Bootstrap 是完全开源的，它的代码托管、开发、维护都依赖________平台。

2．EasyUI 是一种基于________的用户界面插件集合。

二、简答题

1．简述 Bootstrap 前端框架适用的网站类型。

2．简述 EasyUI 前端框架适用的网站类型。

综合实训

实训 1．设计用户类 User，其属性有 id、name、favorites（爱好，数组类型），再自行添加一些其他属性（如性别、民族），请用户注册上述信息，并且在另一个网页上显示用户注册时输入的数据。

实训 2．自学使用 radio 组件。

实训 3．在项目中使用日历组件，如学生注册时添加生日属性。

第 4 章

Spring MVC 框架在项目中的运用

本章学习目标

- 了解 Spring MVC 框架的运行原理
- 掌握 Spring MVC 框架的集成和配置
- 掌握 Spring MVC 框架的请求映射和参数传递
- 熟悉 Spring MVC 校验、格式化和异常处理

本章介绍 Spring MVC 框架的运行原理和使用方法，以及如何将其集成到 Gradle 构建的模块中。本章可分为两个部分，4.1～4.3 节着重讲解 Spring MVC 框架的基本特性，4.4～4.9 节介绍 Spring MVC 框架的高级特性。4.4 节之后的章节会用到数据库连接的知识，建议放在第 5 章之后讲解。

4.1 Spring MVC 运行流程和集成

传统 MVC 模式中，很多应用程序的问题在于处理业务数据的对象和显示业务数据的视图之间存在紧密耦合。通常，请求业务对象的命令都是由视图发起的，因此视图对任何业务对象更改都有高度敏感性。而且，当多个视图依赖于同一个业务对象时，MVC 模式缺乏灵活性。

Spring MVC 的全名为 Spring Web MVC，是 Spring 的一个模块，如图 4-1 所示。它是基于 MVC 设计模式的前端 Web 框架，它使用了 MVC 设计模式的思想，将 Web 层进行职责解耦，实现了 Web MVC 设计模式。它是基于请求驱动类型的轻量级 Web 框架。基于请求驱动指的是使用“请求-响应”模型。使用 Spring MVC 框架会极大地简化日常的 Web 开发流程。

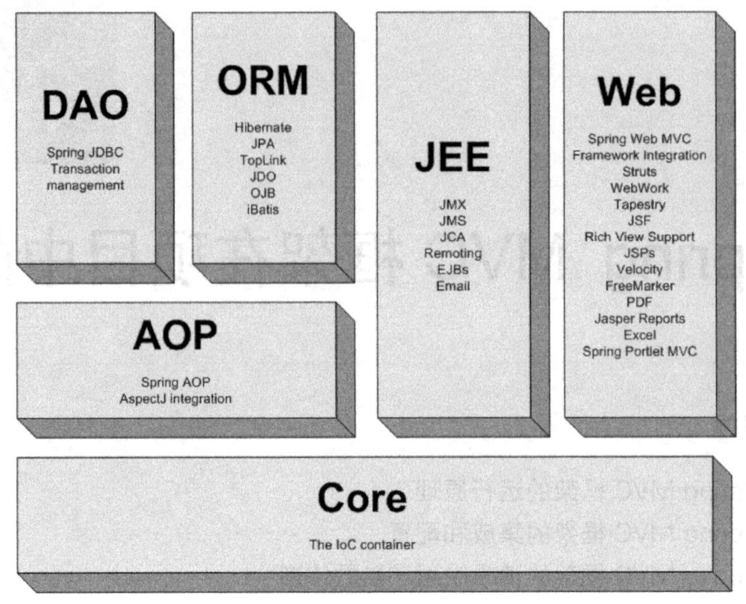

图 4-1　Spring 模块构成

4.1.1　Spring MVC 运行流程

Spring MVC 由前端控制器、上下文、应用控制器和页面控制器（也称为动作）构成，如图 4-2 所示。

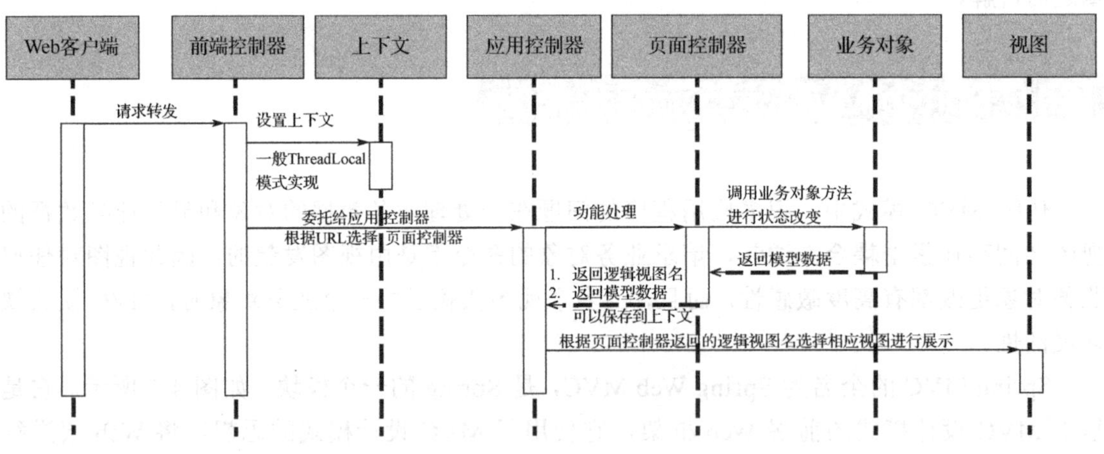

图 4-2　Spring MVC 运行流程图

下面将对各模块的职责进行说明。

前端控制器：负责为表现层提供统一访问点，从而避免出现重复的控制逻辑（由前端控制器统一回调相应的方法）；并且可以为多个请求提供共用的逻辑（如准备上下文等），将选择的具体视图和具体的功能处理分离。

上下文：过去使用 Model 2 模式的时候，为视图准备的要展示的模型数据是直接放在请求中（Servlet API 相关）的。有了上下文之后，就可以将相关数据放置在上下文中，而与协议无关（如 Servlet API）的访问/设置模型数据一般通过 ThreadLocal 模式实现。

应用控制器：前端控制器将选择的具体视图和具体的功能处理分离之后，需要有人来管理，应用控制器作为一种策略设计模式的应用，就是用来选择具体视图技术（视图的管理）和具体的功能处理(页面控制器/命令对象/动作管理)的。它可以很容易地切换视图/页面控制器，使两者相互不产生影响。

页面控制器/动作/处理器：其功能是处理代码，收集参数、封装参数到模型，转调业务对象处理模型，将逻辑视图名返回给前端控制器（和具体的视图技术解耦），由前端控制器委托应用控制器选择具体的视图来展示，可以用命令设计模式实现。页面控制器也称为处理器或动作。

4.1.2 Spring MVC 的核心类和接口

如图 4-3 所示，Spring MVC 框架的核心类 DispatcherServlet，即前端控制器，用于分发和响应用户请求。句柄映射类 HandlerMapping 用于映射用户请求，Controller 类则具体响应用户请求。

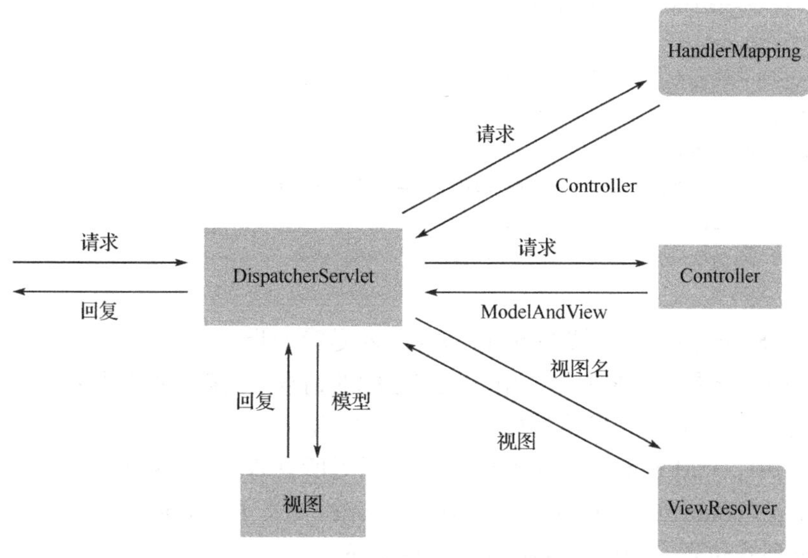

图 4-3 Spring MVC 请求流转图

HandlerMapping—处理 RequestMapping 请求的映射句柄；ModelAndView—模型和视图，使用数据渲染视图；ViewResolver—视图解析器

4.1.3 任务一：项目集成 Spring MVC 框架

要想使用 Spring MVC 框架，就要将 Spring MVC 框架集成到前面基于 Gradle 构建的 Module 中。集成过程有以下三个步骤。

步骤 1．修改 build.gradle 文件，在 dependencies 中添加 Spring MVC 依赖。代码如下：

```
  // https://mvnrepository.com/artifact/org.springframework/spring-web
    compile  group:  'org.springframework',  name:  'spring-web',  version:
'4.3.18.RELEASE'
  // https://mvnrepository.com/artifact/org.springframework/spring-webmvc
```

```
    compile  group:  'org.springframework',  name:  'spring-webmvc',  version: '4.3.18.RELEASE'
```

步骤 2．修改 Spring MVC 在 web.xml 中的配置。代码如下：

```
<?xml version="1.0" encoding="UTF-8"?>
<web-app version="2.5" xmlns="http://java.sun.com/xml/ns/javaee"
  xmlns:xsi="http://www.w3.org/2001/XMLSchema-instance"
  xsi:schemaLocation="http://java.sun.com/xml/ns/javaee
  http://java.sun.com/xml/ns/javaee/web-app_2_5.xsd">
  <!--Springmvc 前端控制器配置-->
  <servlet>
    <servlet-name>dispatcherServlet</servlet-name>
    <servlet-class>org.Springframework.web.servlet.DispatcherServlet</servlet-
    class>
    <init-param>
      <param-name>contextConfigLocation</param-name>
      <param-value>classpath*:spring-mvc.xml</param-value>
    </init-param>
    <load-on-startup>1</load-on-startup>
  </servlet>
  <servlet-mapping>
    <servlet-name>dispatcherServlet</servlet-name>
    <url-pattern>/</url-pattern>
  </servlet-mapping>
</web-app>
```

其中，使用 Spring MVC 配置 DispatcherServlet 是第一步。

DispatcherServlet 是前端控制器，用于拦截匹配的请求（拦截匹配的规则要自己定义），把拦截下来的请求，依据规则分发给目标 Controller 处理。

上面的代码中，<url-pattern>/</url-pattern>表示对所有请求进行拦截。

<param-name>contextConfigLocation</param-name>定义了 Spring MVC 的上下文配置文件 spring-mvc.xml 所在位置，即位于 resources 文件夹下。

步骤 3．在 resources 文件夹下创建 Spring MVC 的配置文件 spring-mvc.xml。文件内容如下：

```
<?xml version="1.0" encoding="UTF-8"?>
<beans xmlns="http://www.springframework.org/schema/beans"
      xmlns:xsi="http://www.w3.org/2001/XMLSchema-instance"
      xmlns:context="http://www.springframework.org/schema/context"
      xmlns:mvc="http://www.springframework.org/schema/mvc"
      xsi:schemaLocation="http://www.springframework.org/schema/beans
http://www.springframework.org/schema/beans/spring-beans.xsd
http://www.springframework.org/schema/context
http://www.springframework.org/schema/context/Spring-context.xsd
http://www.springframework.org/schema/mvc
http://www.springframework.org/schema/mvc/spring-mvc.xsd">
    <!-- 配置自动扫描的包 --> <!-- 自动扫描控制器 -->
    <context:component-scan base-package="com.ssm"/>
```

```
        <!-- 视图渲染 -->
    <bean
    id="internalResourceViewResolver"
class="org.Springframework.web.servlet.view.InternalResourceViewResolver">
            <property name="prefix" value="/WEB-INF/views/"/>
            <property name="suffix" value=".jsp"/>
     </bean>
     <!-- 控制器映射器和控制器适配器 -->
     <mvc:annotation-driven></mvc:annotation-driven>
     <!-- 静态资源映射器 -->
     <mvc:resources mapping="/views/**" location="/WEB-INF/views/" />
    </beans>
```

根据上面的文件，必须要在 Java 源文件目录下创建 com.ssm 包，在 WEB-INF 下创建 views 子目录，否则文件会显示红色的错误提示信息。

提示：如果要集成 Bootstrap 和 EasyUI 前端框架，这里务必在 spring-mvc.xml 文件中添加对它们的静态资源映射，代码如下：

```
<mvc:resources mapping="/bootstrap/**" location="/bootstrap/" />
<mvc:resources mapping="/easyui/**" location="/easyui/" />
```

如果有图片、视频等需要映射的静态资源，也需要在 spring-mvc.xml 文件中添加映射，代码如下：

```
<mvc:resources mapping="/images/**" location="/images/" />
```

4.1.4　Spring MVC 框架控制器中常用的注解说明

spring-mvc.xml 文件中<context:component-scan/> 标签表示 Spring 容器将会扫描指定包中的类的注解，base-package="com.ssm"则定义了扫描的包名。在接下来的章节的代码中将会频繁使用下面这些常用注解：

@Controller：声明控制器组件。

@Service：声明业务类组件。

@Repository：声明存储类组件。

@Component：泛指组件，当不好归为上面三类时使用。

@RequestMapping("/menu")：请求映射。

@Resource：用于注入（J2EE 提供的），默认按名称装配。

@Resource(name="beanName")：用于注入，按 Bean 的名称装配。

@Autowired：用于自动注入（Spring 提供的），默认按类型装配。

@Transactional(rollbackFor={Exception.class})：用于事务管理。

@Scope("prototype")：用于设定 bean 的作用域。

4.1.5　任务二：Spring MVC 的简单实例

按 4.1.3 节集成 Spring MVC 框架后，下面以一个简单的示例演示其使用过程。

【示例 4-1】在 com.ssm.controller 包下创建 StudentController.java 文件，如图 4-4 所示。

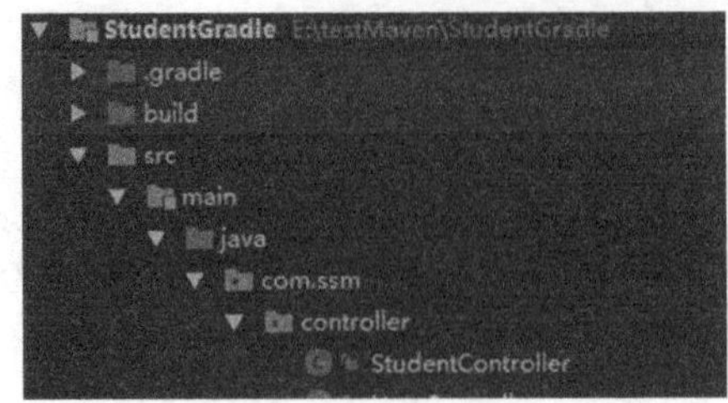

图 4-4　StudentController.java 所在项目结构图

步骤 1．新建 StudentController.java 类，代码如下：

```
@Controller
public class StudentController {
    @RequestMapping("/Student/list.do")
    public String listStudents(){
        return "listStudents";
    }
}
```

上述代码中，@Controller 注解定义 StudentController 类是一个控制器组件的 Bean；方法 listStudents 前面使用了注解@RequestMapping("/Student/list.do")，定义 listStudents 方法的请求映射（也就是浏览器地址栏里的访问形式）是"/Student/list.do"。listStudents 方法体内的 return "listStudents"表示跳转到/WEB-INF/views 下的 listStudents.jsp 页面，如果需要跳转到某个方法，则写成 return "redirect:请求映射"。

步骤 2．在/WEB-INF/views 下创建 listStudents.jsp 文件，在文件中添加字符串“test listStudent.jsp”用于测试。

步骤 3．在浏览器地址栏中输入 http://localhost:8077/Student/list.do（这里的/Student/list.do 在 StudentController 类中使用）。

运行效果如图 4-5 所示。

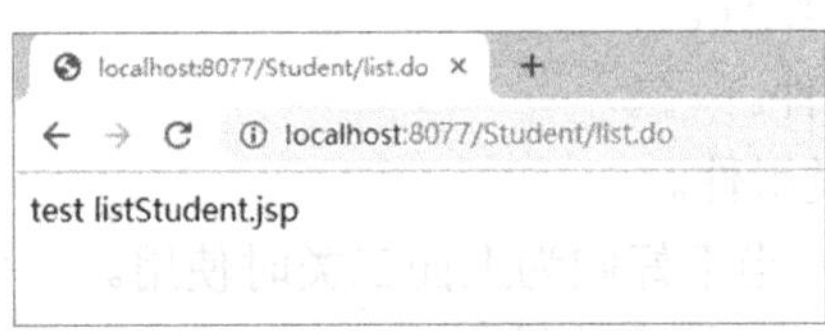

图 4-5　示例 4-1 运行效果图

@RequestMapping("/Student/list.do")定义的请求映射，在 4.2 节中会详细讲述。

拓展练习：设计一个 TeacherController，将请求映射跳转到 listTeachers.jsp 页面。

4.2　Spring MVC 请求映射

@RequestMapping 是 Spring Web 应用程序中常被用到的注解之一。这个注解会将 HTTP 请求映射到 MVC 和 REST 控制器的处理方法上。

4.2.1　@RequestMapping

在 Spring MVC 应用程序中，RequestDispatcher 这个 Servlet 负责将客户端发送过来的 HTTP 请求路由到控制器的处理方法。所有需要配置 Web 请求映射的方法，都必须添加@Request Mapping 注解。

@RequestMapping 注解既可以用在控制器类级别上，也可以用在方法级别上。示例 4-1 就是用在类级别上，而后面示例 4-2 是用在方法级别上的。也可以同时用到类级别和方法级别上，称为分层请求映射。

4.2.2　映射原理

Spring MVC 框架是怎样检查并处理 Controller 上的@RequestMapping 注解的呢？当服务器运行，容器启动时会扫描所有注册的 Bean，依次遍历，判断其是否是处理器，并检测其方法句柄 HandlerMethod，遍历 Handler 中的所有方法，找出其中被@RequestMapping 注解标记过的方法。然后获取方法上的@RequestMapping 实例。再检查方法所属的类有没有@RequestMapping 注解，将类级别的@RequestMapping 和方法级别的@RequestMapping 结合起来，创建请求映射信息表 RequestMappingInfo。当请求到达时，去请求映射信息表 urlMap 中寻找匹配的 URL，并获取对应映射 Mapping 实例，然后到 handlerMethods 中获取匹配 HandlerMethod 的实例。同时，将 RequestMappingInfo 实例及处理器方法注册到缓存中，如图 4-6 所示。

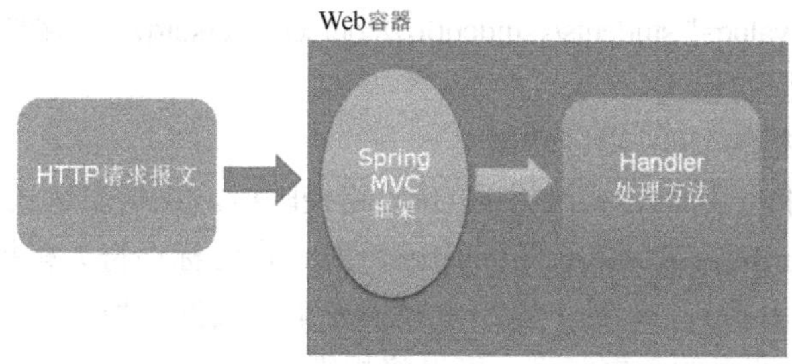

图 4-6　Spring MVC 映射原理

HTTP 请求报文包含 5 部分信息，如图 4-7 所示。

- 请求方法，如 GET 或 POST，表示提交的方式。
- URL，请求的地址信息。
- HTTP 协议及版本。
- 报文头，即请求头信息（包括 Cookie 信息）。
- 报文体，即请求内容区（请求的内容或数据），如表单提交时的参数数据、URL 请求参数（?abc=123?后边的）等。

Spring Web MVC 框架不仅提供了 URL 路径映射，还提供了强大的映射规则。处理器按映射的不同特征，将请求的映射分为如下 4 种：

- URL 路径映射：使用 URL 映射请求到处理器的功能处理方法。
- 请求方法映射限定：如限定功能处理方法只处理 GET 方法的请求。
- 请求参数映射限定：如限定功能处理方法只处理包含“studentId”参数的请求。
- 请求头映射限定：如限定功能处理方法只处理包含“Accept=application/json”的请求。

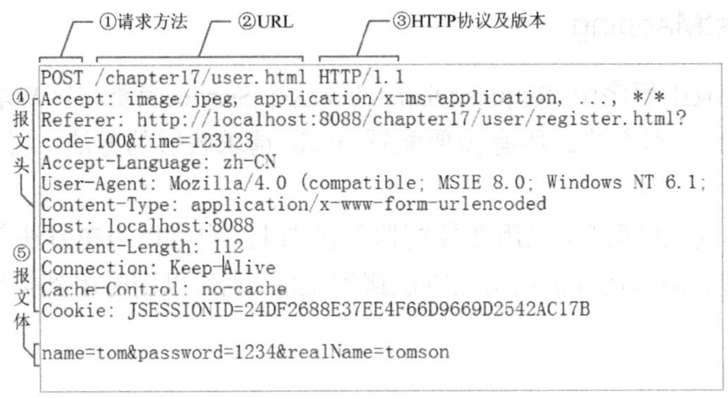

图 4-7　HTTP 请求报文格式

1．普通的 URL 路径映射写法举例

@RequestMapping(value={"/login", "/Student/list.do"})：多个 URL 路径映射到同一个处理器的处理方法。

@RequestMapping(value="/students/{studentId}")：{studentId}为占位符，请求的 URL 可以是 “/students/10341” 或 “/students/abcd” 的形式，通过@PathVariable 提取 URI 模板模式中的{studentId}中的 studentId 变量值。

@RequestMapping(value="/students/{studentId}/create")：请求的 URL 是与“/students/10341/create”类似的形式。

@RequestMapping(value="/students/{studentId}/course/{courseId}")：请求的 URL 是与 “/students/10341/course/101” 类似的形式。

2．正则表达式风格的 URL 路径映射

Spring 3.0 开始支持正则表达式风格的 URL 路径映射。正则表达式风格的 URL 路径映射是一种特殊的 URI（Uniform Resource Identifier，统一资源标识符）模板模式映射。URI 模板模式映射采用{studentId}形式，本身不能指定模板变量的数据类型。但正则表达式风格的 URL 路径映射可以指定模板变量的数据类型，可以将规则写得相当复杂。格式为{变量名:正则表达式}，通过@PathVariable 提取模式中的{变量名:正则表达式匹配的值}中的变量名。下面用实例说明：

@RequestMapping(value="/students/{categoryCode:\\d+}-{pageNumber:\\d+}")：可以匹配 “/students/123-1”，但不能匹配 “/students/abc-1”。

@RequestMapping(value="/students/**")：匹配任意形式，还可以匹配到“/students/abc/abc”，但 “/students/123” 形式将被 URI 模板模式映射中的 “/students/{studentId}” 模式优先映射到。

@RequestMapping(value="/student?")：可匹配“/student1”或“/studenta”，但不匹配“/student”或 “/studentaa”。

@RequestMapping(value="/student*")：可匹配 “studentabc” 或 “/student”，但不匹配 “/studentabc/abc”。

@RequestMapping(value="/student/*")：可匹配 “/student/abc”，但不匹配 “/studentabc”。

@RequestMapping(value="/students/**/{studentId}")：可匹配 “/students/abc/abc/123” 或 “/students/123”，也就是 Ant 风格和 URI 模板变量风格可混用。

4.2.3 任务三：项目中使用分层请求映射

分层请求实际上是指在类级别和方法级别上分别使用注解@RequestMapping，将类的请求和方法的请求分开，以便于代码的阅读。下面的示例中，在类级别上请求使用@RequestMapping("/Student")，在 listStudents 方法级别上请求使用@RequestMapping("list.do")。访问 listStudents 方法的时候，两者要连接起来，即在浏览器地址栏里输入

http://localhost:8077/Student/list.do

如果访问 addStudent 方法，则在地址栏里输入

http://localhost:8077/Student/add.do

```
@Controller
@RequestMapping("/Student")
public class StudentController {
    @RequestMapping("list.do")
    public String listStudents(Model model){
        return "listStudents";
    }
    @RequestMapping("add.do")
    public String addStudent(){
        return "addStudent";
    }
}
```

我们总结通用的写法，如果需要访问 listStudents 方法，则在浏览器地址栏里输入

http://localhost:端口号/项目部署名/Student/list.do

如果需要访问 addStudent 方法，则在浏览器地址栏里输入

http://localhost:端口号/项目部署名/Student/add.do

4.2.4 GET/POST 限定的请求

除了大家熟知的 GET 和 POST 请求方法，标准的方法集合中还包含 PUT、DELETE、HEAD 和 OPTIONS。这些方法的含义连同行为许诺一起在 HTTP 规范中定义。一般浏览器只支持 GET 和 POST 方法。通常情况下，不限定方法时，默认同时支持 GET 和 POST。

【示例 4-2】

```
@RequestMapping(value = "toAdd.do")
public String toAdd(){
    return "addStudent";
}
```

在@RequestMapping 里没有定义请求的方法，则表示支持所有请求，所有 URL 为 <controllerURI>/toAdd.do 的请求都会由 toAdd()处理。

【示例 4-3】在示例 4-2 的基础上，添加 method=RequestMethod.POST 的限定请求，限定 toAdd 方法仅支持 POST 请求。代码如下：

```
@RequestMapping(value = "toAdd.do",method = RequestMethod.POST)
public String toAdd(){
    return "addStudent";
}
```

当再次在浏览器地址栏里输入 toAdd.do 时，会出现如图 4-8 所示的出错提示页面。

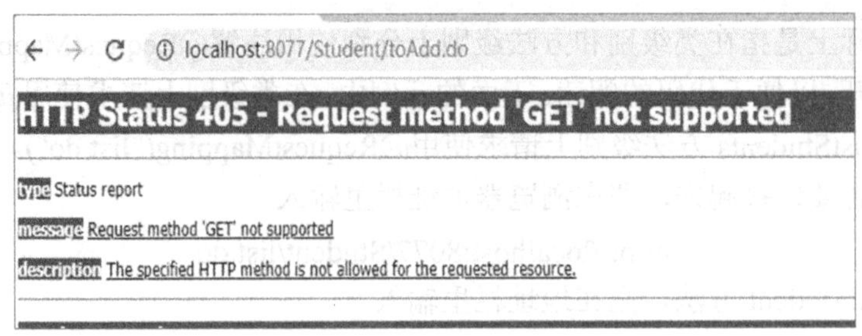

图 4-8　限定请求的出错提示

出错提示表示不支持 GET 方法的访问形式。下面在 index.jsp 页面中使用 POST 方法的请求形式来访问 toAdd 方法。index.jsp 页面的内容如下：

```
<form action="/Student/toAdd.do" method="post">
    <input value="add student" type="submit">
</form>
```

单击 index.jsp 页面中的“提交”按钮，请求以 POST 方法的方式提交给 StudentController 的 toAdd 方法，运行不再出错，正常跳转到 addStudent.jsp 页面，如图 4-9 所示。

图 4-9　运行正常效果

4.3　项目中实现参数传递

传统的 MVC 模式中，Servlet 使用 HttpRequest、HttpResponse 来获取和响应数据。而 Spring MVC 框架简化了控制层，改变了以前将 Servlet 作为中转控制器的方式。那么，使用 Spring MVC 框架，前台的参数是怎样传递到控制层中的呢？控制层又怎样进行页面的跳转呢？本节将就此由简入繁地进行逐步讲解。在下面的行文中，传入指从 JSP 页面传到控制层，传出指从控制层传到 JSP 页面。

4.3.1　任务四：简单参数传入

Spring MVC 框架通过使用注解@RequestParam 传入参数。如果与用户交互时传入页面输入框的参数名称与控制层方法入参名称一致，可以省略注解@RequestParam。下面用两个示例演示简单参数的传入过程。

【示例 4-4】省略注解@RequestParam 示例。用户在 addStudent.jsp 输入框输入 username 和 pwd，单击“提交”按钮，跳转到 StudentController 的 add 方法。下面是 addStudent.jsp 的代码。

```
<form action="add.do" method="post">
    <input name="username" type="text"><br/>
    <input name="pwd" type="password"><br/>
    <input value="add student" type="submit">
</form>
```

在控制层 StudentController 的 add 方法头，定义 username 和 pwd 两个 String 类型的入参，在控制台输出用户输入的 username 和 pwd 变量值，然后跳转到 listStudents.jsp 页面。代码如下：

```
@RequestMapping(value = "add.do",method = RequestMethod.POST)
public String add(String username,String pwd){
    System.out.println(username);
    System.out.println(pwd);
    return "listStudents";
}
```

部署运行，访问 http://localhost:8077/Student/toAdd.do，输入用户名和密码，然后单击 add student 按钮，如图 4-10 所示。控制台输出结果如图 4-11 所示，方框里换行输出 username 和 pwd 的值 admin 和 123。

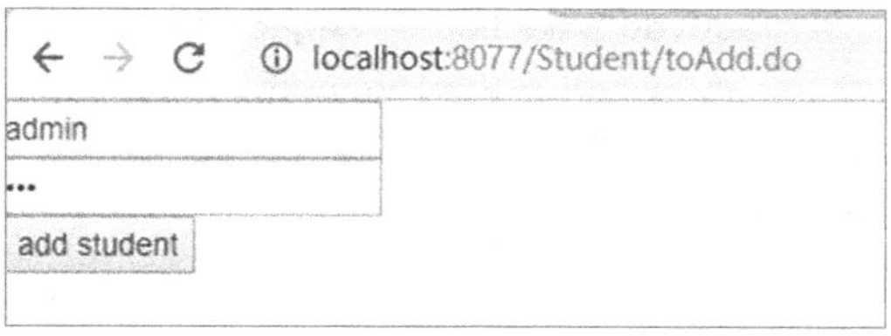

图 4-10　输入用户名和密码

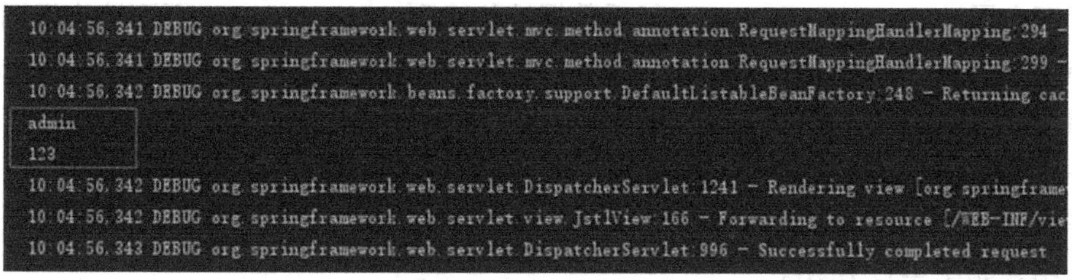

图 4-11　控制台输出结果

【示例 4-5】当与用户交互时传入页面输入框的参数名称与控制层方法入参名称不一致时，在方法入参前使用注解@RequestParam。代码如下：

```
@RequestMapping(value = "add.do",method = RequestMethod.POST)
public String add(@RequestParam("username") String name, @RequestParam("pwd")
String password){
    System.out.println(name);
    System.out.println(password);
    return "listStudents";
}
```

在示例 4-4 代码的基础上，分别在参数前加上@RequestParam("username")和 @RequestParam("pwd")。

很明显，添加@RequestParam 注解会使代码显得啰嗦，并且再次命名容易混淆参数，所以在编写代码时尽量不自找麻烦，保持入参名称和用户交互页面输入框中参数名称一致。

4.3.2 任务五：简单数据传出

Spring MVC 框架使用 Model 或者 ModelAndView 对象把数据从控制层传到 JSP 渲染视图。Model 只是用来传输数据的，并不会进行业务的寻址。ModelAndView 除了能传输数据，还可以进行业务寻址，就是能手工设置要渲染的网页。

两者还有一个最大的区别，即 Model 的每一次请求都由 Spring MVC 框架自动创建对象，而 ModelAndView 则需要用户创建对象。

还有一个对象 ModelMap 可以用于传输数据，它继承自 LinkedHashMap。对于 ModelMap 的请求，Spring MVC 框架自动创建实例并作为控制层方法的入参，用户无须自己创建。这里不做更为详细的介绍。大家可自行上网学习其使用方法。

【示例 4-6】在示例 4-4 的基础上，使用 Model 对象传出 username 和 pwd 的值来渲染 listStudents.jsp 页面。修改 StudentController.java 的 add 方法，代码如下：

```
@RequestMapping(value = "add.do",method = RequestMethod.POST)
public String add( String username, String pwd,Model model){
    model.addAttribute("username",username);
    model.addAttribute("pwd",pwd);
    return "listStudents";
}
```

在上面的 add 方法的入参处添加了 Model 对象，并在方法体中使用 model.addAttribute 方法添加了两个属性 username 和 pwd，然后数据被传输到 listStudents.jsp 页面渲染视图。

对 listStudents.jsp 页面稍做修改，增加如下两个表达式，使其显示 username 和 pwd 两个参数的值。

```
${username}<br/>
${pwd}
```

运行结果如图 4-12 所示。

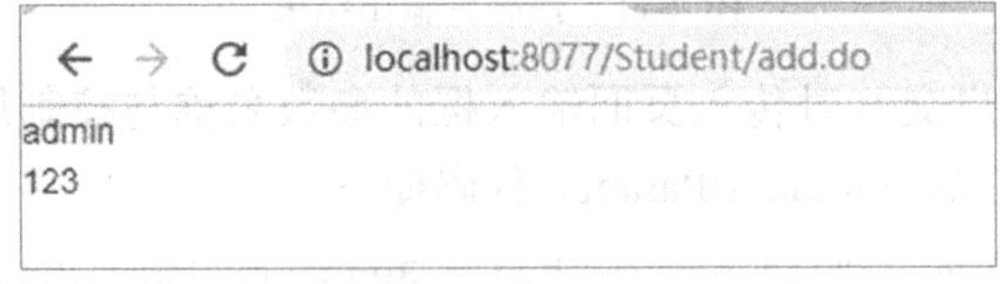

图 4-12　示例 4-6 运行结果

【示例 4-7】使用 ModelAndView 对象来实现数据的传出，代码如下：

```
@RequestMapping(value = "add.do",method = RequestMethod.POST)
public ModelAndView add( String username, String pwd){
    ModelAndView mv=new ModelAndView();
    mv.addObject("username",username);
    mv.addObject("pwd",pwd);
    mv.setViewName("listStudents");
```

```
        return mv;
    }
```

首先，add 方法的返回类型从 String 改成了 ModelAndView，然后在方法体里创建了一个 ModelAndView 的对象 mv，然后给 mv 对象添加了 username 和 pwd 两个变量对象，并设置了视图 listStudents.jsp，方法结束时返回对象 mv。运行结果跟图 4-12 一致，说明数据已经传出并渲染了视图。

4.3.3 任务六：实体对象参数传递

当要传递的参数很多时，使用参数逐个传递的方法就很麻烦。通常来说，实际项目中，一般会使用客户端传递整个对象。下面以 Student 类为例来示范实体对象作为参数的传入。

【示例 4-8】实体对象参数传递。

步骤 1. 在 com.ssm.model 包下创建 Student.java 类，为其创建三个属性：name、pwd、age，以及实现它们的 getter/setter 方法。代码如下：

```
package com.ssm.model;

/**
 * Created by Julia on 2019-8-7.
 */
public class Student {
    private String name;
    private String pwd;
    private int age;

    public String getName() {
        return name;
    }

    public void setName(String name) {
        this.name = name;
    }

    public String getPwd() {
        return pwd;
    }

    public void setPwd(String pwd) {
        this.pwd = pwd;
    }

    public int getAge() {
        return age;
    }

    public void setAge(int age) {
        this.age = age;
    }
}
```

步骤 2．对 addStudent.jsp 的代码稍做修改，三个输入框的 name 属性值必须跟 Student 类的属性名称一一对应，如下面的 name、pwd、age。代码如下：

```
<form action="add.do" method="post">
    <input name="name" type="text"><br/>
    <input name="pwd" type="password"><br/>
    <input name="age" type="text"><br/>
    <input value="add student" type="submit">
</form>
```

步骤 3．StudentController.java 中 add 方法传递参数的代码也要修改。参数改为 Student 类型的对象 student，ModelAndView 添加的对象也要同步改为 student 对象，如图 4-13 所示。

```
@RequestMapping(value = "add.do", method = RequestMethod.POST)
public ModelAndView add(Student student) {
    ModelAndView mv=new ModelAndView();
    mv.addObject("student", student);
    mv.setViewName("listStudents");
    return mv;
}
```

图 4-13　实体对象作为参数传入代码

步骤 4．listStudents.jsp 源码中，为所有表达式的变量名称都加上 student 对象，后面跟随属性名称，代码如下：

```
${student.name}<br/>
${student.pwd}<br/>
${student.age}
```

步骤 5．部署运行，再次访问 http://localhost:8077/Student/add.do，运行结果如图 4-14 所示。

localhost:8077/Student/add.do
admin
123
12

图 4-14　示例 4-8 运行结果

4.3.4　任务七：Cookie 值传递

Cookie 这种小型文本文件，是网站为了辨别用户身份，进行 Session 跟踪而储存在用户本地终端上的数据（通常经过加密），是由用户客户端计算机暂时或永久保存的信息。它由服务器创建，然后添加到 HttpServletResponse 中并发送给客户端（浏览器），之后浏览器的每次请求（HttpServletRequest）里都会携带“Cookie 数组”。Cookie 由键值名和键值组成。相同域名和路径下的键值名不能重复，如果添加键值名重名的键值对，则会覆盖上一个同名的键值对。添加 Cookie 时要指定 Cookie 所在域（setPath），指定存在时长（setMaxAge）。代码中可以同时添加多个 Cookie 键值对。

Spring MVC 可以通过两种方式获取 Cookie 值：

（1）通过 HttpServletRequest 中的 getCookies 方法获取 Cookie 数组，然后迭代里面的每一个 Cookie 键值对。

（2）在控制器中通过注解@CookieValue（键值名），获取某个指定的 Cookie。

下面通过示例 4-9 和示例 4-10 分别演示上述两种取值方式。

【示例 4-9】使用 HttpServlet Request 中的 getCookies 方法获取 Cookie 值。

步骤 1．修改 StudentController.java 中的 add 方法，入参里添加 HttpServletResponse 类型的对象 res，在方法体里新建 Cookie 类型的键值为<city:zhuhai>的 cookie，并添加到 res 对象里。代码如下：

```
@RequestMapping(value = "add.do",method = RequestMethod.POST)
    public ModelAndView add(Student student, HttpServletResponse res){
        Cookie cookie=new Cookie("city","zhuhai");
        res.addCookie(cookie);
        ModelAndView mv=new ModelAndView();
        mv.addObject("student",student);
        mv.setViewName("listStudents");
        return mv;
}
```

步骤 2．修改 listStudents.jsp 代码，添加一个代码段，获取 request 隐式对象的数组 cookies。如果数组 cookies 不为空，则循环迭代输出数组的每一个键值对。代码如下：

```
<%
    Cookie[] cookies = request.getCookies();    //根据请求数据，找到 cookie 数组
    if (null==cookies) {                        //如果没有 cookie 数组
        System.out.println("没有 cookie");
    } else {
        for(Cookie cookie : cookies){
                out.println("<br/>"+cookie.getName()+":"+cookie.getValue());
        }
    }
%>
```

步骤 3．部署运行，再次访问 http://localhost:8077/Student/add.do，运行结果如图 4-15 所示。

图 4-15　示例 4-9 运行结果

【示例 4-10】使用注解@CookieValue 获取 Cookie 值。

步骤 1．修改 StudentController.java 中的 add 方法，使其定位到 list.do 请求。代码如下：

```
@RequestMapping(value = "add.do",method = RequestMethod.POST)
    public ModelAndView add(Student student, HttpServletResponse res){
        Cookie cookie=new Cookie("city","zhuhai");
```

```
        res.addCookie(cookie);
        ModelAndView mv=new ModelAndView();
        mv.addObject("student",student);
        mv.setViewName("redirect:list.do");
        return mv;
    }
```

步骤 2．修改 StudentController.java 中的 list 方法，入参里添加注解@CookieValue，获取名为 city 的 Cookie 值，输出到控制台上。代码如下：

```
    @RequestMapping("list.do")
    public String list(@CookieValue("city")String city){
        System.out.println(city);
        return "listStudents";
    }
```

步骤 3．部署运行，再次访问 http://localhost:8077/Student/add.do，控制台输出 zhuhai，如图 4-16 所示。

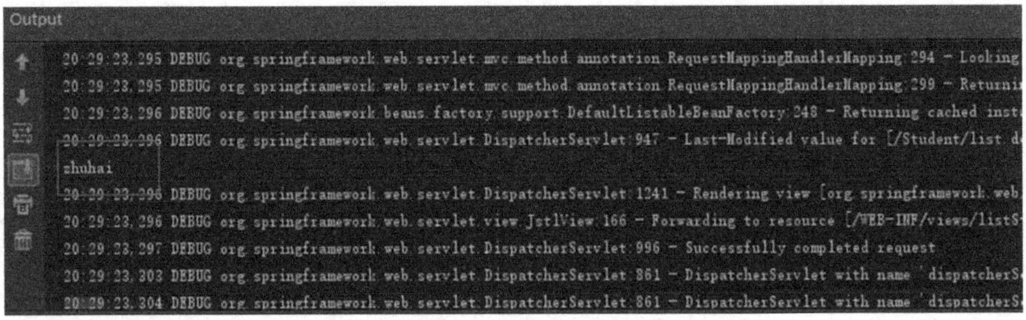

```
Output
20:29:23,295 DEBUG org.springframework.web.servlet.mvc.method.annotation.RequestMappingHandlerMapping:294 - Looking
20:29:23,295 DEBUG org.springframework.web.servlet.mvc.method.annotation.RequestMappingHandlerMapping:299 - Returni
20:29:23,296 DEBUG org.springframework.beans.factory.support.DefaultListableBeanFactory:248 - Returning cached inst
20:29:23,296 DEBUG org.springframework.web.servlet.DispatcherServlet:947 - Last-Modified value for [/Student/list.d
zhuhai
20:29:23,296 DEBUG org.springframework.web.servlet.DispatcherServlet:1241 - Rendering view [org.springframework.web.
20:29:23,296 DEBUG org.springframework.web.servlet.view.JstlView:166 - Forwarding to resource [/WEB-INF/views/listS
20:29:23,297 DEBUG org.springframework.web.servlet.DispatcherServlet:996 - Successfully completed request
20:29:23,303 DEBUG org.springframework.web.servlet.DispatcherServlet:861 - DispatcherServlet with name 'dispatcherS
20:29:23,304 DEBUG org.springframework.web.servlet.DispatcherServlet:861 - DispatcherServlet with name 'dispatcherS
```

图 4-16 控制台输出结果

4.3.5 任务八：Session 值传递

在计算机中，尤其是在网络应用中，Session 被称为“会话控制”。Session 对象用来存储用户会话所需的属性及配置信息。这样，当用户在 Web 应用程序的网页之间跳转时，存储在 Session 对象中的变量将不会丢失，而是在整个用户会话中一直存在下去。当用户请求来自应用程序的网页时，如果该用户还没有会话，则 Web 服务器将自动创建一个 Session 对象。只有当会话过期或被主动放弃时，服务器才终止该会话。

Session 会话机制是一种服务器端机制，它使用类似于哈希表的结构来保存信息。服务器根据客户端的请求（HttpServletRequest）创建 Session（request.getSession()）。Session 存储在服务器端，每一个 Session 都有一个 ID。当创建一个 Session 后，会将该 Session 的 sessionID 存放到此次访问的 Cookie 中。当下次客户端的访问到来，需要提取服务器中的 Session 时，会根据访问的 Cookie 里的 sessionID 值来找到服务器中的具体 Session。服务器会把长时间没有活动的 Session 从服务器内存中清除，此时 Session 便失效。Tomcat 服务器中 Session 的默认失效时间为 30 分钟。这个时长可以根据需要修改。

默认情况下，Spring MVC 框架会将模型中的数据存储到请求域中。当一个请求结束后，数据就失效了。如果要跨页面使用，就要用到 Session 会话存储数据。使用注解@SessionAttributes

可以使模型中的数据存储到会话域中。它有三个参数需要定义：

names：字符串数组，填写需要存储到 Session 中的数据的名称。

types：根据指定参数的类型，将模型中对应类型的参数存储到 Session 中。

value：和 names 是一样的。

跟 Cookie 一样，Session 类型数据的存取过程也有两种方式。第一种使用 HttpSession 类型的对象，第二种使用注解@SessionAttributes。下面分别用示例 4-11 和示例 4-12 来演示 Session 数据的存取过程。

【示例 4-11】使用 HttpSession 类型对象获得 Session 类型数据。

步骤 1．修改 StudentController.java 文件中的 add 方法，在入参里定义 HttpSession 类型的变量 session，在方法体内为 session 对象添加属性，属性名为 student，属性值是传进来的入参 student。代码如下：

```
@RequestMapping(value = "add.do",method = RequestMethod.POST)
   public ModelAndView add(Student student, HttpSession session){
       session.setAttribute("student",student);
       ModelAndView mv=new ModelAndView();
       mv.addObject("student",student);
       mv.setViewName("redirect:list.do");
       return mv;
   }
```

步骤 2．修改 StudentController.java 文件中的 list 方法，在入参里定义 HttpSession 类型的变量 session，在方法体内使用 getAttribute 方法获取上面定义过的 student 属性，并赋值给 Student 类型的变量 s1，然后在控制台输出 s1 的名称。代码如下：

```
@RequestMapping("list.do")
   public String list(HttpSession session){
       Student s1= (Student) session.getAttribute("student");
       System.out.println(s1.getName());
       return "listStudents";
}
```

步骤 3．部署运行，访问 http://localhost:8077/Student/toAdd.do，进入 addStudent.jsp 页面，在“姓名”输入框输入 admin，如图 4-17 所示。

图 4-17　输入姓名

单击 add student 按钮，控制台输出结果如图 4-18 所示，控制台上输出了变量 s1 的名称 admin。

图 4-18　控制台输出结果

提示：*在 JSP 或者控制层中使用 HttpResponse 或者 HttpSession 前，必须在 build.gradle 文件中添加两个依赖，请参考示例 4-11 中源码文件的 build.gradle 文件内容：*

```
// https://mvnrepository.com/artifact/javax.servlet/javax.servlet-api
   compile group: 'javax.servlet', name: 'javax.servlet-api', version: '3.1.0'
// https://mvnrepository.com/artifact/javax.servlet/jstl
   compile group: 'javax.servlet', name: 'jstl', version: '1.2'
```

【示例 4-12】使用注解@SessionAttributes 获取 Session 类型数据。

沿用示例 4-11 的代码，修改 StudentController.java 中的 add 方法，将 HttpSession 类型的入参改为 Model 类型。代码如下：

```
@Controller
@RequestMapping("/Student")
public class StudentController {
    @RequestMapping("list.do")
    public String list(HttpSession session){
        Student s1= (Student) session.getAttribute("student");
        System.out.println(s1.getName());
        return "listStudents";
    }
    @RequestMapping(value = "toAdd.do")
    public String toAdd(){
         return "addStudent";
    }
    @RequestMapping(value = "add.do",method = RequestMethod.POST)
    public ModelAndView add(Student student,Model model){
        model.addAttribute("student",student);
        ModelAndView mv=new ModelAndView();
        mv.addObject("student",student);
        mv.setViewName("redirect:list.do");
        return mv;
    }
}
```

按上面代码部署运行，控制台会在 `System.out.println(s1.getName());` 处报空指针异常，因为 student 对象是通过 Model 添加的，只在请求时有效。如图 4-19 所示，我们使用注解

@SessionAttributes 来改正，在类前面的@RequestMapping 后增加语句@SessionAttributes ("student")，将 student 对象的生命有效期变为会话域有效。修改后，再按照示例 4-12 的步骤部署运行，控制台上会成功输出 admin。

提示：如果要在 JSP 页面上获取 Session 类型数据，可以使用如下方式。

```
${sessionScope.student.username}，也可以直接写成${student.name}
```

本节完整地讲述了 Spring MVC 传递数据的方式，涵盖了简单类型数据、对象数据及 Cookie、Session 类型数据。在实际应用中，请大家根据实际情况灵活运用注解和非注解的形式。

```
@Controller
@RequestMapping("/Student")
@SessionAttributes("student")
public class StudentController {


    @RequestMapping("list.do")
    public String list(HttpSession session){
        Student s1= (Student) session.getAttribute( name: "student");
        System.out.println(s1.getName());
        return "listStudents";
    }
    @RequestMapping(value = "toAdd.do")
    public String toAdd() { return "addStudent"; }
    @RequestMapping(value = "add.do",method = RequestMethod.POST)
    public ModelAndView add(Student student,Model model){

        model.addAttribute("student",student);
        ModelAndView mv=new ModelAndView();
        mv.addObject("student",student);
        mv.setViewName("redirect:list.do");
        return mv;
    }
}
```

图 4-19　使用@SessionAttributes 获取 Session 类型数据

4.4 项目中的数据格式化

4.4.1 Spring MVC 框架的格式化

Spring MVC 框架自带格式化数据功能，可以在 JSP 页面上把数据变成我们想要的样式。Spring MVC 框架使用 AnnotationFormatterFactory<A extends Annotation>来格式化数据。其中包括两个重要注解：

@DateTimeFormat：日期时间类型。

@NumberFormat：数字类型，包括货币类型、正常数字类型、百分数类型。

将注解作用于 POJO 的属性后，页面传递的字符串将会转换成对应的格式化数据。在软件开发中强烈建议使用注解的方式。

在下面的代码中，对 4 个属性分别使用不同的格式化注解，供大家学习其使用方式。

```
public class Student implements Serializable {
    //日期时间类型
   @DateTimeFormat(pattern = "yyyy-MM-dd")
    private Date birthday;
    //正常数字类型
   @NumberFormat(style = NumberFormat.Style.NUMBER,pattern = "#,###")
    private int total;
    //百分数类型
    @NumberFormat(style = NumberFormat.Style.PERCENT)
    private double discount;
    //货币类型
    @NumberFormat(style = NumberFormat.Style.CURRENCY)
    private double money;
}
```

上面代码中实现了以下功能。

① 格式化 birthday 为"yyyy-MM-dd"的日期时间类型。

② 格式化 total 为每三位用逗号隔开的数字类型。

③ 格式化 discount 为百分数类型。

④ 格式化 money 为当前区域的货币类型。

4.4.2 任务九：使用 Spring MVC 的数据格式化功能

【示例 4-13】使用注解@NumberFormat 格式化用户输入的数据。

步骤 1．在 Student 类中添加属性 discount 及其 getter/setter 方法，对属性 discount 使用格式化注解@NumberFormat，将其格式化为百分数类型。代码如下：

```
//百分数类型
    @NumberFormat(style = NumberFormat.Style.PERCENT)
    private double discount;
    public double getDiscount() {
        return discount;
    }
    public void setDiscount(double discount) {
        this.discount = discount;
}
```

步骤 2．在 addStudent.jsp 页面添加 discount 属性的输入框，代码如下：

```
折扣：<input name="discount" type="text"><br/>
```

步骤 3．修改 listStudents.jsp 页面，主要代码如下：

```
<%@ taglib prefix="spring" uri="http://www.springframework.org/tags" %>
...
<spring:eval expression="student.discount"/><br/>
${student.name}<br/>
${student.pwd}<br/>
${student.age}
```

提示：代码里使用了 Spring 标签库中的 eval 标签来输出 student.discount 属性值，只有这个标签才能正确地格式化数据。

步骤 4．部署运行，在 addStudent.jsp 页面中输入折扣 0.8，如图 4-20 所示。

单击 add Student 按钮，页面跳转到 listStudents.jsp，显示折扣为 80%，数据格式化成功，如图 4-21 所示。

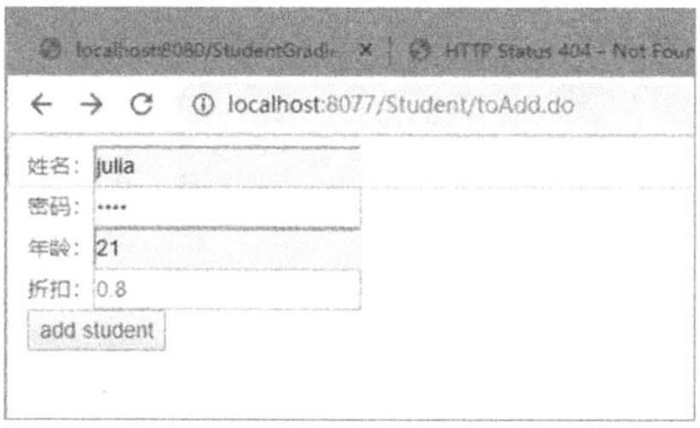

图 4-20 在 addStudent.jsp 页面输入折扣 0.8

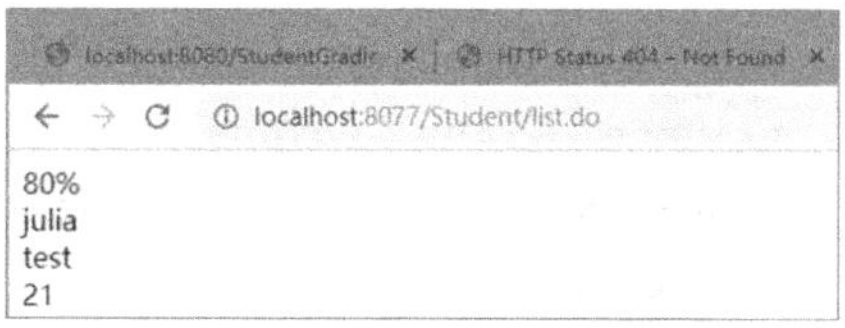

图 4-21 折扣格式化结果

4.5 项目中使用服务器端校验

4.5.1 Spring MVC 的服务器端校验

输入校验分为客户端校验和服务器端校验，客户端校验主要用于过滤正常用户的误操作，通常通过 JS 完成。客户端校验能把用户误操作的非法输入阻止在客户端，减少服务器的负载压力。服务器端校验是整个 Web 应用程序阻止非法数据的最后防线，主要通过编程实现。对于用户的恶意行为，客户端校验无能为力，必须通过服务器端校验才能防止恶意侵入。因此，在 Web 应用程序的设计中，客户端校验和服务器端校验必须同时具备。

Java EE 企业级应用技术的 Bean Validation 为 JavaBean 的验证定义了相应的元数据模型和 API。默认的元数据是 Java Annotations，通过使用 XML 可以对原有的元数据信息进行覆盖和扩展。在应用程序中，通过使用 Bean Validation 或是自定义的约束，如@NotNull、@Length、@ZipCode，可以确保数据模型（JavaBean）的正确性。约束既可以附加到字段，也可以附加到 getter 方法、类或者接口上面。对于一些特定的需求，用户可以自定义约束。Bean Validation 是一个运行时的数据验证框架，在验证之后验证的错误信息会被马上返回。

Bean Validation 1.0 定义了 Java EE 企业级应用技术的 JSR 303 规范。JSR 是 Java Specification Requests 的缩写，意思是 Java 规范提案，它向 JCP（Java Community Process）提出了新增一个标准化技术规范的正式请求。任何人都可以提交 JSR，以向 Java 平台增添新的 API 和服务。JSR 已成为 Java 中的一个重要标准。JSR 303 规范是 Java EE 6 中的一项子规范，专门定义实体 Bean 的校验规范。

Hibernate Validator 框架就是 Bean Validation 规范的参考实现。它提供了 JSR 303 规范中所有内置约束的实现，除此之外还有一些附加的约束，其应用十分广泛。

Spring MVC 框架没有服务器端校验的功能，它使用 Hibernate Validator 进行数据校验。这里就使用 Hibernate Validator 框架来进行服务器端校验。在使用前，需要在 build.gradle 文件里添加 hibernate-validator 依赖，代码如下：

```
// https://mvnrepository.com/artifact/org.hibernate/hibernate-validator
    compile  group:  'org.hibernate',  name:  'hibernate-validator',  version: '5.4.1.Final'
```

在下面的代码中，在每个属性前面都添加一个用于校验的注解。校验规则请参考注解前的注释。

```
public class Student implements Serializable {
    //校验是否为空
    @NotBlank(message = "登录名不能为空")
    private String loginname;
    @NotBlank(message = "密码不能为空")
    //校验长度
    @Length(min = 6,max = 8,message = "密码长度必须在 6 到 8 位之间")
    private String password;
    @NotBlank(message = "用户名不能为空")
    private String username;
    //校验数值范围
    @Range(min = 15,max = 60,message = "年龄必须在 15 到 60 岁之间")
    private int age;
    //校验是否是合法的邮箱
    @Email(message = "必须是合法的邮箱地址")
    private String email;
    @DateTimeFormat(pattern = "yyyy-MM-dd")
    //校验生日是否是已经过去的日期
    @Past(message = "生日必须是过去的一个日期")
    private Date birthday;
    //使用正则表达式校验电话号码的格式
    @Pattern(regexp = "[1][3,8][3,6,9][0-9]{8}",message = "无效的电话号码")
    private String phone;
}
```

Hibernate Validator 框架完整的校验注解，请参看表 4-1。

表 4-1　校验注解表

校验注解	校验数据类型	说明
@AssertFalse	Boolean	校验注解的元素值是 false
@AssertTrue	Boolean	校验注解的元素值是 true
@NotNull	任意类型	校验注解的元素值不是 null
@Null	任意类型	校验注解的元素值是 null
@Min(value=值)	BigDecimal、Biginteger、byte、short、int、long 等任何 Number 或 CharSequence（存储的是数字）子类型	校验注解的元素值大于或等于@Min 指定的 value 值
@Max(value=值)	和@Min 要求一样	校验注解的元素值小于或等于@Max 指定的 value 值

续表

验证注解	验证数据类型	说明
@DecimalMin(value=值)	和@Min 要求一样	校验注解的元素值大于或等于@DecimalMin 指定的 value 值
@DecimalMax(value=值)	和@Min 要求一样	校验注解的元素值小于或等于@DecimalMax 指定的 value 值
@Digits(integer=整数位数，fraction=小数位数)	和@Min 要求一样	校验注解的元素值的整数位数和小数位数上限
@Size(min=下限、max=上限)	字符串、Collection、Map、数组等	校验注解的元素值在 min 和 max（包含）指定区间之内，如字符长度、集合大小
@Past	java.util.Date, java.util.Calendar, Joda Time 类库的日期类型	校验注解的元素值（日期类型）比当前时间早
@Future	与@Past 要求一样	校验注解的元素值（日期类型）比当前时间晚
@NotBlank	CharSequence 子类型	校验注解的元素值不为空（不为 null，去除首位空格后长度不为 0），不同于@NotEmpty，@NotBlank 只应用于字符串且在比较时会去除字符串的首位空格
@Length(min=下限，max=上限)	CharSequence 子类型	校验注解的元素值长度在 min 和 max 区间内
@NotEmpty	CharSequence 子类型、Collection、Map、数组	校验注解的元素值不为 null 且不为空（字符串长度不为 0，集合大小不为 0）
@Range(min=最小值，max=最大值)	BigDecimal、BigInteger、CharSequence、byte、short、int、long 等原子类型和包装类型	校验注解的元素值在最小值和最大值之间
@Email(regexp=正则表达式，flag=标志的模式)	CharSequence 子类型（如 String）	校验注解的元素值是 email，也可以通过 regexp 和 flag 指定自定义的 email 格式
@Pattern(regexp=正则表达式，flag=标志的模式)	String 和任何 CharSequence 的子类型	校验注解的元素值与指定的正则表达式匹配
@Valid	任何非原子类型	指定递归校验关联的对象； 如用户对象中有地址对象属性，且想在校验用户对象时一起校验地址对象，在地址对象上加@Valid 注解即可级联校验

4.5.2 任务十：项目中实现 Spring MVC 的服务器端校验

下面通过示例 4-14 演示如何使用注解完成校验。

【示例 4-14】

步骤 1．在 build.gradle 文件添加 hibernate-validator 依赖，代码如下：

```
// https://mvnrepository.com/artifact/org.hibernate/hibernate-validator
compile group: 'org.hibernate', name: 'hibernate-validator', version: '5.4.1.Final'
```

步骤 2．修改 resources/spring-mvc.xml 文件内容，添加校验器 validator 的 Bean。它由 Spring

框架的 LocalValidatorFactoryBean 工厂创建，注入属性是 Hibernate Validator 框架的 HibernateValidator 类。然后使用标签 mvc:annotation-driven 定义了 validator 校验器是注解驱动的。代码如下：

```
<!-- 校验器，配置 validator -->
    <bean  id="validator"  class="org.Springframework.validation.beanvalidation.
LocalValidatorFactoryBean">
        <property  name="providerClass"  value="org.hibernate.validator.Hibernate
Validator"></property>
    </bean>
    <!-- 控制器映射器和控制器适配器 -->
    <mvc:annotation-driven  validator="validator"></mvc:annotation-driven>
```

步骤 3．修改 com.ssm.entity/Student.java 源文件，在属性 name 前添加不为空和长度 3～20 字符的校验规则。message 定义了出错后的返回信息。代码如下：

```
 @NotNull(message = "姓名不能为空")
@Size(min = 3,max = 20,message = "长度不合适")
private String name;
```

步骤 4．修改 StudentController.java 源文件，给入参 student 添加@Validated 注解，并且紧随其后必须添加 BindingResult 对象，否则将无法显示出错信息。代码如下：

```
public String add(@Validated Student student, BindingResult result){
        if(result.hasErrors()){
        List<ObjectError> errors=result.getAllErrors();
        for (ObjectError objectError:errors){
            System.out.println(objectError);
        }
        return "addStudent";
    }
    else{studentServiceImpl.add(student);
    return "redirect:list.do";}
}
```

步骤 5．测试校验。部署运行，访问 http://localhost:8077/Student/toAdd.do。在 addStudent.jsp 页面，故意在姓名输入框里只输入两个字符，单击 add student 按钮。控制台将输出 name 属性值被拒绝的出错信息，依然停留在 addStudent.jsp 页面。详细的错误信息如下：

```
Field error in object 'student' on field 'name': rejected value [ju]; codes
[Size.student.name,Size.name,Size.java.lang.String,Size];        arguments
[org.Springframework.context.support.DefaultMessageSourceResolvable:    codes
[student.name,name]; arguments []; default message [name],20,3]; default message [姓
名长度必须是 3~20 字符]
```

4.6 Spring MVC 上传

4.6.1 Spring MVC 上传的实现类

Spring MVC 框架默认的解析器里没有对文件上传的解析，所以，在用解析器处理文件上

传时需要用 Spring MVC 提供的 MultipartResolver 类。

当收到文件上传请求时，DispatcherServlet 的 checkMultipart()方法会调用 MultipartResolver 的 isMultipart()方法判断请求中是否包含文件。如果请求数据中包含文件，则调用 MultipartResolver 的 resolveMultipart()方法对请求的数据进行解析，然后将文件数据解析成 MultipartFile 并封装在 MultipartHttpServletRequest（继承了 HttpServletRequest）对象中，最后传递给控制器。

MultipartFile 封装了请求数据中的文件，此时这个文件存储在内存中或临时的磁盘中，因为请求结束后临时存储空间将被清空，所以需要将其转存到一个合适的位置。在 MultipartFile 接口中有如下方法：

- String getName()：获取参数的名称。
- String getOriginalFilename()：获取文件的原名称。
- String getContentType()：获取文件内容的类型。
- boolean isEmpty()：判断文件是否为空。
- long getSize()：获取文件大小。
- byte[] getBytes()：将文件内容以字节数组的形式返回。
- InputStream getInputStream()：将文件内容以输入流的形式返回。
- void transferTo(File dest)：将文件内容传输到指定文件中。

因为 StandardServletMultipartResolver 实现了 MultipartResolver 接口，所以我们可以在 spring-mvc.xml 配置文件中使用 StandardServletMultipartResolver 类来定义 Bean。

4.6.2　任务十一：对项目实现上传功能

下面在示例 4-15 中演示如何使用 StandardServletMultipartResolver 实现文件上传。

【示例 4-15】

步骤 1. 修改配置文件 spring-mvc.xml，添加 StandardServletMultipartResolver 的 bean 对象 multipartResolver。代码如下：

```
    <bean  id="multipartResolver"  class="org.Springframework.web.multipart.support.
StandardServletMultipartResolver">
    </bean>
```

步骤 2. 修改 webapp/WEB-INF/web.xml，给 DispatcherServlet 添加 multipart-config 标签元素。如果出现红色错误提示：this element is not allowed here，不用理会。另外，multipart-config 的子标签<location>d:/upload</location>，表示本例在 d 盘上新建了 upload 文件夹，用于存放上传的文件。其他标签的意义请参看代码注释。代码如下：

```
    <!--Springmvc 前端控制器配置-->
    <servlet>
    <servlet-name>dispatcherServlet</servlet-name>
    <servlet-class>org.Springframework.web.servlet.DispatcherServlet</servlet-class>
          <init-param>
              <param-name>contextConfigLocation</param-name>
              <param-value>classpath:Spring-mvc.xml</param-value>
          </init-param>
```

```
        <load-on-startup>1</load-on-startup>
        <multipart-config>
            <!--上传到 d:/upload 目录-->
            <location>d:/upload</location>
            <!--文件大小为 2MB-->
            <max-file-size>2097152</max-file-size>
            <!--整个请求不超过 4MB-->
            <max-request-size>4194304</max-request-size>
            <!--所有文件都要写入磁盘-->
            <file-size-threshold>0</file-size-threshold>
        </multipart-config>
    </servlet>
```

步骤 3．修改 Student.java 源文件，为其添加属性 image 及其 getter/setter 方法，用于存放图片的路径。代码如下：

```
Private String image;
//getter/setter 方法
```

步骤 4．修改上传表单 addStudent.jsp，添加一个文本框 file 用于选择要上传的文件。值得注意的是，要修改 form 标签，添加 enctype="multipart/form-data"属性，表示该表单将会提交多媒体文件。具体代码如下：

```
    <form:form action="add.do" method="post" class="form-control" enctype="multipart/
form-data">
```

另外，添加文件选择控件，代码如下：

```
    <div class="form-group">
    <label for="file" class="col-sm-2 control-label">头像：</label>
    <div class="col-sm-10">
        <input type="file" class="form-control" id="file" name="file" placeholder=
"head image">
    </div>
  </div>
```

步骤 5．修改 StudentController.java 源文件中的 add 方法，添加一个 MultipartFile 类型的入参 file，调用 uploadFile 方法实现上传，并将该文件的完整路径赋值给 Student 的 file 属性。代码如下：

```
   public String add(@RequestParam(value = "file")MultipartFile file,  Student student)
throws IOException {
   ...
            uploadFile(file);
            student.setImage(file.getOriginalFilename());
            model.addAttribute("student", student);
            return "listStudents";
   ...
   }
```

下面是用于文件上传的 uploadFile 方法的具体实现代码，通过调用 File 类的 transferTo 方法实现了文件上传。

```
private void uploadFile(MultipartFile image) throws IOException{
    if(image!=null){
        // 上传后的文件保存目录及名称
        File imageFile = new File(image.getOriginalFilename());
        image.transferTo(imageFile);// 将上传文件保存到相应位置
    }
}
```

步骤 6．部署运行，访问 http://localhost:8077/Student/toAdd.do。在 addStudent.jsp 页面输入各种属性，如图 4-22 所示。这里需要选择一个文件用于上传，单击 add student 按钮后，检查 d:/upload 目录，观察是否存在上传的图片。

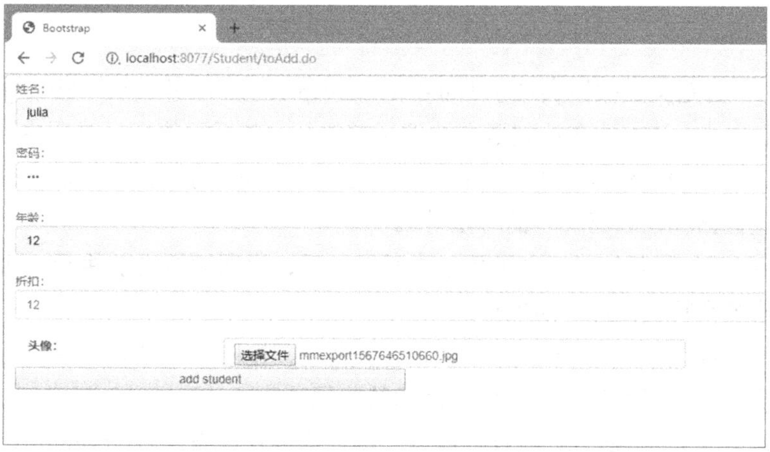

图 4-22　上传图片

4.7　Spring MVC 拦截器

4.7.1　拦截器的定义

Spring MVC 中的拦截器（Interceptor）类似于 Servlet 中的过滤器（Filter），它主要用于拦截用户请求并做相应的处理。通过拦截器可以进行权限验证、记录请求信息的日志、判断用户是否登录等。

使用 Spring MVC 中的拦截器前，需要对拦截器类进行定义和配置。通常拦截器类可以通过以下两种方式来定义。

（1）通过实现 HandlerInterceptor 接口，或继承 HandlerInterceptor 接口的实现类（如 HandlerInterceptorAdapter）来定义。

（2）通过实现 WebRequestInterceptor 接口，或继承 WebRequestInterceptor 接口的实现类来定义。

下面以实现 HandlerInterceptor 接口的方式为例，自定义拦截器类的代码如下：

```
public class CustomInterceptor implements HandlerInterceptor{
    public boolean preHandle(HttpServletRequest request,
```

```
                            HttpServletResponse response, Object handler)
                            throws Exception {
        return false;
    }
    public void postHandle(HttpServletRequest request,
                          HttpServletResponse response, Object handler,
                          ModelAndView modelAndView) throws Exception {

    }
    public void afterCompletion(HttpServletRequest request,
                               HttpServletResponse response, Object handler,
                               Exception ex) throws Exception {

    }
}
```

上述代码中，自定义拦截器实现了 HandlerInterceptor 接口，并实现了接口中的三个方法：

preHandle()方法：该方法在控制器方法调用前执行，其返回值表示是否中断后续操作。当其返回值为 true 时，表示继续向下执行；当其返回值为 false 时，会中断后续的所有操作（包括调用下一个拦截器和控制器类中的方法等）。

postHandle()方法：该方法在控制器方法调用之后、视图解析之前执行。可以通过此方法对请求域中的模型和视图做进一步的修改。

afterCompletion()方法：该方法会在整个请求完成、视图渲染结束之后执行。可以通过此方法实现资源清理、内存回收、记录日志信息的收尾工作。

开发拦截器就像开发 Servlet 或者过滤器一样，需要在配置文件 spring-mvc.xml 中进行配置，代码如下：

```
<!--配置拦截器-->
    <mvc:interceptors>
        <!--<bean class="com.ssm.interceptor.CustomeInterceptor" />-->
        <!--拦截器 1-->
        <mvc:interceptor>
            <!--配置拦截器的作用路径-->
            <mvc:mapping path="/**"/>
            <mvc:exclude-mapping path=""/>
            <!--定义在<mvc:interceptor>下面的表示对匹配指定路径的请求才进行拦截-->
            <bean class="com.ssm.interceptor.Intercptor1"/>
        </mvc:interceptor>
        <!--拦截器 2-->
        <mvc:interceptor>
            <mvc:mapping path="/hello"/>
            <bean class="com.ssm.interceptor.Interceptor2"/>
        </mvc:interceptor>
```

上面的代码中，<mvc:interceptors>元素用于配置一组拦截器，其子元素<bean>中定义的是全局拦截器，它会拦截所有请求；而<mvc:interceptor>元素中定义的是指定路径的拦截器，它只对指定路径下的请求生效。<mvc:interceptor>元素的子元素<mvc:mapping>用于配置拦截器作用的路径，该路径在其属性 path 中定义。在上述代码中，path 的属性值“/**”表示拦截所有路径，“/hello”表示拦截所有以“/hello”结尾的路径。如果在请求路径中排除不需要拦截的内容，还可以通过<mvc:exclude-mapping>元素进行配置。

注 意 <mvc:interceptor>中的子元素必须按照上述代码中的配置顺序进行编写，即<mvc:mapping> <mvc:exclude-mapping> <bean>，否则文件会报错。

4.7.2 任务十二：对项目实现拦截器功能

下面将在登录场景下，使用 Spring MVC 的拦截器功能实现用户的安全检测。

【示例 4-16】功能：只允许登录过的用户访问/WEB-INF/views 下的文件，否则返回登录页面。

步骤 1．检查 web.xml，确认 dispatcherServlet 前端控制器将对*.do 访问形式的请求进行分发。代码如下：

```
<servlet-mapping>
      <servlet-name>dispatcherServlet</servlet-name>
      <url-pattern>*.do</url-pattern>
</servlet-mapping>
```

步骤 2．创建包 inter，在其下创建新的源文件 MyInterceptor.java。MyInterceptor 类实现了 HandlerInterceptor 接口及其 preHandle、postHandle、afterHandle 三个方法。只有 preHandle 方法有具体实现代码，获取 Session 类型变量 username，如果为空则表明尚未登录到 login.jsp 页面，否则正常往下执行业务。代码如下：

```
package inter;
import org.Springframework.web.servlet.HandlerInterceptor;
import org.Springframework.web.servlet.ModelAndView;
import javax.servlet.http.HttpServletRequest;
import javax.servlet.http.HttpServletResponse;
import javax.servlet.http.HttpSession;
public class MyInterceptor implements HandlerInterceptor {
   public boolean preHandle(HttpServletRequest request, HttpServletResponse
   response, Object handler) throws Exception {
      HttpSession session=request.getSession();
      String username= (String) session.getAttribute("username");
      if (username==null)
      { request.getRequestDispatcher("/login.jsp").forward(request,response);
         return  false;}
      else
      {return true;}
   }
   public void preHandle(HttpServletRequest request, HttpServletResponse response,
   Object handler, ModelAndView modelAndView) throws Exception {
   }
   public void preHandle(HttpServletRequest request, HttpServletResponse response,
   Object handler, Exception ex) throws Exception {
   }
}
```

步骤 3．修改 resources/spring-mvc.xml 文件，添加拦截器的配置信息，使用 inter.MyInterceptor 作为拦截器，拦截所有/Student 和/Teacher 的 URL 访问请求。具体代码如下：

```
<mvc:interceptors>
        <mvc:interceptor>
            <mvc:mapping path="/Student/*"/>
            <mvc:mapping path="/Teacher/*"/>
            <bean class="inter.MyInterceptor"/>
        </mvc:interceptor>
</mvc:interceptors>
```

步骤 4．部署运行，在未登录的情况下尝试访问 http://localhost:8077/Student/toAdd.do，结果进入 login.jsp 页面。说明拦截器起到了作用，达到了拦截的目的，如图 4-23 所示。

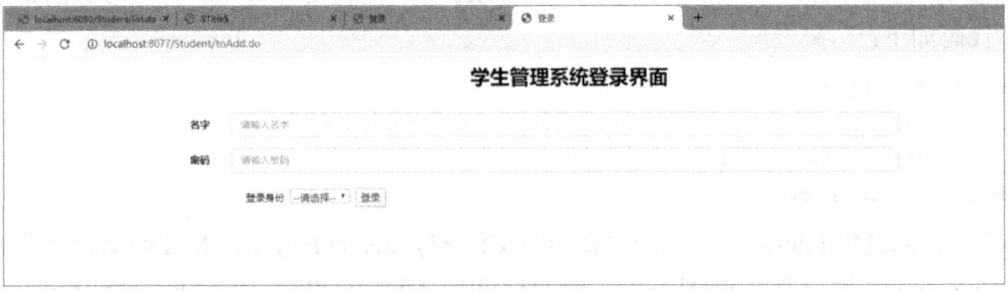

图 4-23　拦截效果图

4.8　Spring MVC 异常处理

在 Java Web 项目的开发中，无论是数据库访问层的操作过程、业务层的调用过程，还是控制层的控制过程，都难免会遇到各种可预知的或不可预知的异常。如果每个方法或者类都需要单独处理异常，系统的代码耦合度会变得很高，开发工作量大且不好统一，维护的工作量也很大。

那么，能不能将所有类型的异常处理从各处理过程中解耦出来独立进行？这样既保证了相关处理过程的功能较单一，也实现了异常信息的统一处理和维护。答案是肯定的。下面将介绍使用 Spring MVC 框架统一处理异常的实现过程。另外，Spring MVC 框架还有三种自定义异常的处理办法，在 4.8.3 节中会有详细的介绍。

4.8.1　全局性系统异常的处理方法

当系统发生异常时，用户会面对 404、500 等服务器内部的页面，目前大多数服务器支持在 web.xml 中通过<error-page>(Websphere/Weblogic)或者<error-code>(Tomcat)节点配置特定异常情况的显示页面。修改 web.xml 文件，增加以下代码：

```
 <!-- 出错页面定义 -->
<error-page>
    <exception-type>java.lang.Throwable</exception-type>
    <location>/500.jsp</location>
```

```
    </error-page>
    <error-page>
        <error-code>500</error-code>
        <location>/500.jsp</location>
    </error-page>
    <error-page>
        <error-code>404</error-code>
        <location>/404.jsp</location>
    </error-page>
```

上面的代码定义了三种异常的处理方法。第一个 error-page 定义了如果出现 Throwable 类型的系统异常，网页将自动跳转到 500.jsp 页面。第二个 error-page 定义了如果出现 500 的系统异常，网页也将自动跳转到 500.jsp 页面。第三个 error-page 则定义了如果出现 404 的系统异常，网页将跳转到 404.jsp 页面。

拓展测试：在浏览器地址栏中输入一个不存在的 URL，系统会自动跳转到自定义的 404.jsp 页面。

4.8.2 任务十三：项目中使用简单异常处理器 SimpleMappingExceptionResolver

SimpleMappingExceptionResolver 实现了 HandlerExceptionResolver 接口，该接口定义了一个 resolveException 方法。要处理异常，需要实现这个接口类，并且实现 resolveException 方法，在 resolveException 方法中处理异常逻辑。而 SimpleMappingExceptionResolver 只能处理简单的异常，当发生异常的时候，会根据发生的异常类型跳转到指定的页面来显示异常信息。下面使用示例 4-17、示例 4-18 来演示 SimpleMappingExceptionResolver 的处理过程和效果。

【示例 4-17】系统异常处理。

步骤 1. 修改 resources/spring-mvc.xml 文件，添加异常处理类 SimpleMappingExceptionResolver 创建的 simpleMappingExceptionResolver。在 bean 的子标签里定义 exceptionMappings 属性。该属性里定义了自定义异常 DatabaseException、InvalidCreditCardException 及对应的出错视图 databaseError.jsp 和 creditCardError.jsp。后面则定义了默认的出错处理视图 error.jsp，异常的属性名为 ex。代码如下：

```
    <bean id="simpleMappingExceptionResolver" class="org.Springframework.web.servlet.
handler.SimpleMappingExceptionResolver">
        <property name="exceptionMappings">
            <map>
                <!-- key:异常类别（非全称），视图名称 -->
                <entry key="DatabaseException" value="databaseError"/>
                <entry key="InvalidCreditCardException" value="creditCardError"/>
            </map>
        </property>
        <!-- 默认的错误处理页面，异常的名称 -->
        <property name="defaultErrorView" value="error"/>
        <property name="exceptionAttribute" value="ex"/>
    </bean>
```

步骤 2. 新增默认的异常处理文件 WEB-INF/views/error.jsp，其内容如下：

```
Sorry,error! ${ex.message }
```

步骤 3．设计一个系统错误，引发异常处理。创建一个实现登录的 LoginController.java 源文件，登录方法里加一条语句 int i = 3/0，引发除数为 0 的异常。具体内容如下：

```
@Controller
public class LoginController {
    @RequestMapping("check.do")
    public String check(String username,String password){
        int i=3/0;
        return "login";
    }
}
```

步骤 4．部署运行，访问 http://localhost:8077/login.jsp，单击“登录”按钮，网页上输出 Sorry,error! / by zero，如图 4-24 所示。

图 4-24　示例 4-17 运行结果

【示例 4-18】自定义异常处理。

步骤 1．在控制器下设计一个自定义异常类 DatabaseException，集成父类 Exception。

```
public class DatabaseException extends Exception {}
}
```

步骤 2．在 WEB-INF/views 下新建文件 databaseError.jsp，代码如下：

```
database error
```

步骤 3．修改 LoginController.java 源文件，删掉 int i = 3/0，加一个条件判断语句，如果用户输入的 username 是 database，则抛出自定义异常 DatabaseException，跳转到 login.jsp 文件。代码如下：

```
@Controller
public class LoginController {
    @RequestMapping("check.do")
    public String check(String username,String password) throws DatabaseException
    {
        if(username.equals("database"))
            throw new DatabaseException();
        return "redirect:/login.jsp";
    }
}
```

步骤 4．部署运行，访问 http://localhost:8077/login.jsp，在用户名输入框中输入 database，单击“登录”按钮，跳转到 databaseError.jsp 页面，网页上显示 database error，如图 4-25 所示。

图 4-25　示例 4-18 运行结果

4.8.3　Spring MVC 自定义异常处理的三种方式

1. 使用@ ExceptionHandler 注解实现异常处理

首先要自定义一个名字重复异常类 NameDuplicateException，定义其无参和有参的构造方法皆继承自父类 Exception，代码如下：

```
public class NameDuplicateException extends Exception {
    public NameDuplicateException() {
        super();
    }
    public NameDuplicateException(String message) {
        super(message);
    }
}
```

然后在需要进行异常处理的控制器里添加异常处理方法，代码如下：

```
@ExceptionHandler(NameDuplicateException.class)
    public String exception(NameDuplicateException e){
        e.printStackTrace();
        return "nameDuplicate";
    }
```

使用该注解不好的地方就是，进行异常处理的方法必须与出错的方法在同一个控制器里面，不能实现全局控制异常，每个类都要重写一遍异常处理方法，并且对代码有很不规范的侵入行为。

2. 实现 HandlerExceptionResolver 接口来自定义异常处理器

（1）在某个用于测试的控制器里添加名字重复异常，并抛出异常，部分代码如下：

```
if(username.equals("test")) {
    throw new NameDuplicateException("username duplicate");
}
```

（2）定义名字重复异常类。代码如下：

```
public class NameDuplicateException extends Exception {
    public NameDuplicateException() {
        super();
    }
    public NameDuplicateException(String message) {
        super(message);
    }
}
```

（3）实现自定义异常处理类 MyException，它实现了 HandlerExceptionResolver 接口，实

现了 resolveException 方法。方法体里首先在控制台输出异常信息，然后定义了异常的出错处理视图是 nameDuplicate.jsp。当出现异常请求而跳转到该视图时会携带异常对象 ex。代码如下：

```
@Component  //定义异常处理类为一个组件
 public class MyException implements HandlerExceptionResolver {
   public ModelAndView resolveException(HttpServletRequest request,
   HttpServletResponse response, Object handler,Exception ex) {
       System.out.println("This is exception handler method!"+ex.getMessage());
       ModelAndView mav = new ModelAndView();
       mav.addObject("ex", ex);
       mav.setViewName("nameDuplicate");
       return mav;
   }
}
```

3．使用@ControllerAdvice 实现全局异常处理

前面说到使用@ExceptionHandler 注解实现异常处理时进行异常处理的方法必须与出错的方法在同一个控制器里。当代码中加入@ControllerAdvice 时，则不需要二者必须在同一个控制器中了。这也是 Spring 3.2 的新特性。也就是说，@ControllerAdvice + @ ExceptionHandler 也可以实现全局的异常捕捉。异常处理类的具体代码如下：

```
@ControllerAdvice  //定义这是一个增强的 controller
 public class MyException implements HandlerExceptionResolver {
@ExceptionHandler(Exception.class)
   public ModelAndView resolveException(Exception ex) {
       System.out.println("This is exception handler  method!"+ex.getMessage());
       ModelAndView mav = new ModelAndView();
       mav.addObject("ex", ex);
       mav.setViewName("nameDuplicate");
       return mav;
   }
}
```

使用@ControllerAdvice 实现全局异常处理，只需要定义类。在该类中，可以定义多个方法，不同的方法处理不同的异常，如专门处理 404 异常的方法、专门处理 500 异常的方法；也可以直接像上面代码一样，在一个方法中处理所有的异常信息。

@ExceptionHandler 注解用来指明异常的处理类型，即如果这里指定为 NameDuplicaterException，则其他异常处理就不会调用这个方法。

4.9 Spring MVC 处理国际化

4.9.1 Spring MVC 框架国际化简介

Spring MVC 中实现国际化的类有两种：ResourceBundleMessageSource 和 ReloadableResourceBundleMessageSource，后者可以把资源文件放到任何位置，不需要重启就能加载文件，并且

可以设置刷新时间。对项目实现国际化有基于浏览器、基于 Session、基于 Cookie 和基于 URL 四种形式。这里只探讨基于浏览器形式的，实际上，其他形式也都是基于 ReloadableResource BundleMessageSource，只是配置有所增加。

4.9.2　任务十四：项目实现国际化

【示例 4-19】对项目实现国际化。

步骤 1．修改 resources/spring-mvc.xml 文件，添加如下代码：

```
<bean id="messageSource" class="org.springframework.context.support.Reloadable
ResourceBundleMessageSource">
        <!-- 国际化信息所在的文件名,根据 ResourceBundleMessageSource 类加载资源文件.\src\
        main\resources\messages\messages_zh_CN.properties -->
        <property name="basename" value="classpath:messages/messages" />
        <!-- 如果在国际化资源文件中找不到对应代码的信息，就用这个代码作为名称  -->
        <property name="useCodeAsDefaultMessage" value="true" />
    </bean>
    <!-- 存储区域设置信息 -->
    <bean id="localeResolver" class="org.springframework.web.servlet.i18n.Session
    LocaleResolver"/>
     <!--配置国际化拦截器-->
    <mvc:interceptors>
        <mvc:interceptor>
            <mvc:mapping path="/**" />
            <bean class="org.springframework.web.servlet.i18n.LocaleChange Interceptor">
                <property name="paramName" value="lang" />
            </bean>
        </mvc:interceptor>
    </mvc:interceptors>
```

上面代码定义了国际化属性文件名是 messages 包下的 messages 文件。

步骤 2．在 resources 包下创建子包 messages，然后创建 messages_zh_CN.properties、messages_en_US.properties。代码如下：

```
messages_zh_CN.properties:
title= 开启
username= 用户名
messages_en_US.properties:
username=username
title=begin to start
```

步骤 3. 在 StudentController.java 源文件中添加一个引导到 views/index.jsp 文件的方法，代码如下：

```
@RequestMapping(value="/index")
        public String  test(HttpServletRequest request, Model model){
            return "index";
        }
```

提示：必须从控制器进入 JSP 页面，否则国际化信息不会渲染 JSP 文件。

步骤 4．在 WEB-INF/views 下创建 index.jsp 文件，代码如下：

```
<%@ page language="java" contentType="text/html; charset=UTF-8"
      pageEncoding="UTF-8"%>
<%@ taglib uri="http://www.springframework.org/tags" prefix="spring"%>
<!DOCTYPE html PUBLIC "-//W3C//DTD HTML 4.01 Transitional//EN" "http://www.w3.
org/TR/html4/loose.dtd">
<html>
<head>
   <meta http-equiv="Content-Type" content="text/html; charset=UTF-8">
   <title>international</title>
</head>
<body>
<div class="login">
   <h1><spring:message code="title" /> </h1>
   <form action="test.do" method="post">
      <input type="text" name="name" placeholder=<spring:message code="username"
      /> required="required" value="" />
   </form>
</div>
</body>
</html>
```

上面代码中，<%@ taglib uri="http://www.springframework.org/tags" prefix="spring"%>定义了使用 Spring 的标签库。<spring:message code="title" /> 表示将从国际化资源属性文件中查找名称为 title 的属性所对应的值。

步骤 5．在 webapp 根目录下创建新文件 show.jsp，实现不同语言的属性文件的选择。代码如下：

```
<%@ page language="java" contentType="text/html; charset=UTF-8"
      pageEncoding="UTF-8"%>
<%@taglib prefix="mvc" uri="http://www.springframework.org/tags/form" %>
<%@taglib prefix="spring" uri="http://www.springframework.org/tags" %>
<html>
<head>
   <title>I18N</title>
</head>
<body>
<a href="/Student/index?lang=zh_CN">中文</a><br>
<a href="/Student/index?lang=en_US">English</a>
</body>
</html>
```

步骤 6．部署运行，访问 http://localhost:8080/show.jsp，显示效果如图 4-26 所示。

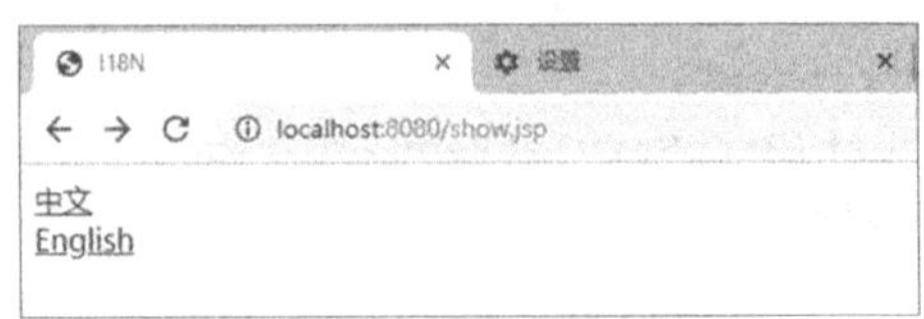

图 4-26　show.jsp 页面显示效果

分别单击“中文”超链接和 English 超链接，显示效果如图 4-27 和图 4-28 所示。

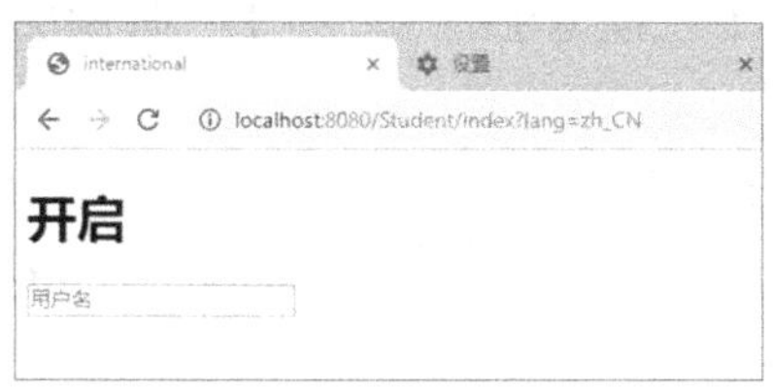

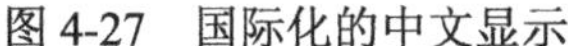
图 4-27　国际化的中文显示

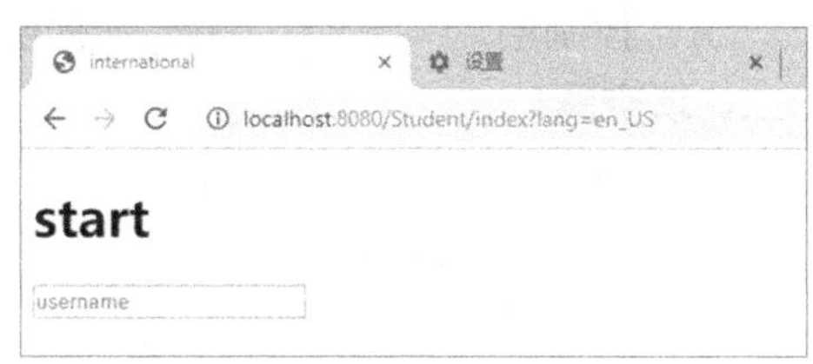

图 4-28　国际化的英文显示

小　结

Spring MVC 框架是 Spring 框架的一部分，它实现了基于 Java 的 Web MVC 模式的轻量级框架。它采用了架构模式的思想，将 Web 层解耦，极大地减轻了开发人员的工作量，并规范了开发流程。本章通过大量案例详细阐述了 Spring MVC 框架常用的基础特性，如请求映射、参数传递，以及高级特性，如数据格式化、服务器端校验和异常处理等。

除了学会使用上述特性完成 Web 层代码的编写和流程的控制，同学们还要学会如何在前面几章的基于 Gradle 构建的项目中集成该框架。在接下来的章节中，还会讲述 Spring MVC 框架与其他框架的集成，配置文件会更为复杂。

习　题

一、单选题

1．关于 Spring MVC，以下说法错误的是（　　）。

A．Spring MVC 的核心入口是 DispatcherServlet

B．@RequestMapping 注解既可以用在类上，也可以用在方法上

C．@PathVariable 的作用是取出 URL 中的模板变量作为参数

D．控制器默认是单例，通过@Scope(“prototype”)注解改为多例，成员变量共享

2．关于 Spring MVC 的核心控制器 DispatcherServlet 的作用，以下说法错误的是（　　）。

A．它负责接收 HTTP 请求

B．加载配置文件

C．实现业务操作

D．初始化上下文应用对象 ApplicationContext

3．Spring MVC 的核心入口类是（　　）。

A．DispatcherServlet　　B．ActionServlet

C．StrutsPrepareAndExecuteFilter　　D．Servlet

4．Spring MVC 中把视图和数据都合并在一起的类是（　　）。

A．View　　B．ModelAndView　　C．DataView　　D．ViewResolver

5．下列（　）不属于 MVC 架构。

A．控制层　　B．数据层　　C．模型层　　D．视图层

二、填空题

1．Spring MVC 框架的核心处理类________，也叫前端控制器。

2．________是 Spring Web 应用程序中常被用到的注解之一。这个注解将 HTTP 请求映射到 MVC 和 REST 控制器的处理方法上。

3．Spring MVC 框架传入参数，使用注解________。

4．Spring MVC 中的拦截器（Interceptor）类似于 Servlet 中的________。

5．输入校验分为________和________。

三、简答题

1．Spring MVC 框架在三层结构里处于什么位置？

2．请描述 Spring MVC 中拦截器的使用场景。

综合实训

实训 1．使用 ArrayList 实现 Student 对象的添加和数据的显示。

实训 2．使用 Spring MVC 实现实体对象数据的传出。

需完成下列功能：

① 创建实体对象 Teacher，属性有 id、name、course。

② 使用户能添加 Teacher 对象，如题图 4-1 所示。

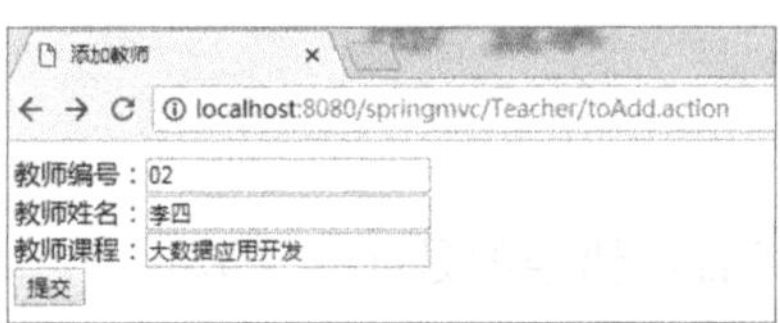

题图 4-1

③ 将添加的 Teacher 值用 List 列表的形式显示，如题图 4-2 所示。

题图 4-2

实训 3．实现 4.6 节中上传图片的下载功能。

实训 4．为实训 2 添加国际化支持，至少 3 个语言版本。

第 5 章

MyBatis 框架在项目中的运用

本章学习目标

- 了解 MyBatis 框架的运行原理
- 掌握 MyBatis 框架的集成和配置
- 掌握 MyBatis 框架对数据库的各种操作
- 熟悉数据库连接池技术

本章将会介绍 MyBatis 框架如何在 IDEA 平台下使用 MyBatis Generator 工具自动生成持久层，以及该框架与第 4 章项目集成的具体过程。除此之外，还会重点讲解如何在 Web 项目中使用 MyBatis 框架对数据库进行各种简单和复杂的操作。

5.1 MyBatis 框架介绍

Java 本身提供了数据库操作的 JDBC 库和基础类，但程序员使用起来存在如下问题：

（1）数据库连接创建、释放频繁造成系统资源浪费，从而影响系统性能。如果使用数据库连接池，可解决此问题。

（2）SQL 语句在代码中存在硬编码，造成代码不易维护。实际应用中 SQL 语句变化的可能性较大，SQL 语句变动后需要修改 Java 代码。

（3）使用 PreparedStatement 向占位符传参数存在硬编码，因为 SQL 语句的 where 条件不确定，可能多也可能少，修改 SQL 语句还要修改代码，系统不易维护。

（4）对结果集解析存在硬编码（查询列名），SQL 语句变化导致解析代码变化，系统不易维护，如果能将数据库记录封装成实体对象，则解析比较方便。

为了改变上述情况，一些优秀的持久层框架出现了，如 iBATIS（MyBatis 前身）和 Hibernate。它们帮我们把那些复杂的重复性工作进行了抽象，从业务处理逻辑中剥离，使得我们不需要再写那些我们不想写但必须写的代码。

MyBatis 本是 Apache 的开源项目 iBATIS，2010 年这个项目由 Apache Software Foundation 迁移到 Google Code，并且改名为 MyBatis，2013 年 11 月迁移到 GitHub。

iBATIS 一词来源于“internet”和“abatis”的组合，是一个基于 Java 的持久层框架。iBATIS 提供的持久层框架包括 SQL Maps 和 Data Access Objects（DAO）。

MyBatis 是一个支持普通 SQL 查询、存储过程和高级映射的优秀持久层框架，它消除了几乎所有的 JDBC 代码和参数的手工设置及对结果集的检索封装，使开发者只需要关注 SQL 本身，而不需要花费精力去处理注册驱动、创建连接、创建语句、手动设置参数、结果集检索等 JDBC 繁杂的过程代码。MyBatis 可以将简单的 XML 或注解用于配置和原始映射，将接口和 Java 的 POJO（Plain Old Java Objects，普通的 Java 对象）映射成数据库中的记录。

MyBatis 可以通过 XML 或注解两种方式将要执行的各种 Statement（statement、preparedStatemnt、CallableStatement）配置起来，并通过 Java 对象和 statement 中的动态参数进行映射，生成最终执行的 SQL 语句，最后由 MyBatis 框架执行 SQL 语句并将结果映射成 Java 对象并返回。

Hibernate 框架也曾经是盛极一时的持久层框架，但 MyBatis 框架和 Hibernate 框架比较起来，有如下特征和优势：

（1）如果有超过千万级别的表，有单次业务大批量数据（百万条及以上）提交需求，或者有单次业务大批量数据（百万条及以上）读取需求，则最好选用 MyBatis 框架。

（2）如果主要业务表的关联表超过 20 个（大概数量，根据表的大小不同有差异），建议使用 MyBatis 框架。

（3）如果多数开发成员不常使用 Hibernate，建议使用 MyBatis 框架。

除了了解 MyBatis 框架的源起，还要从全局上对 MyBatis 框架的架构有一个比较清晰的认识。MyBatis 框架的架构如图 5-1 所示。

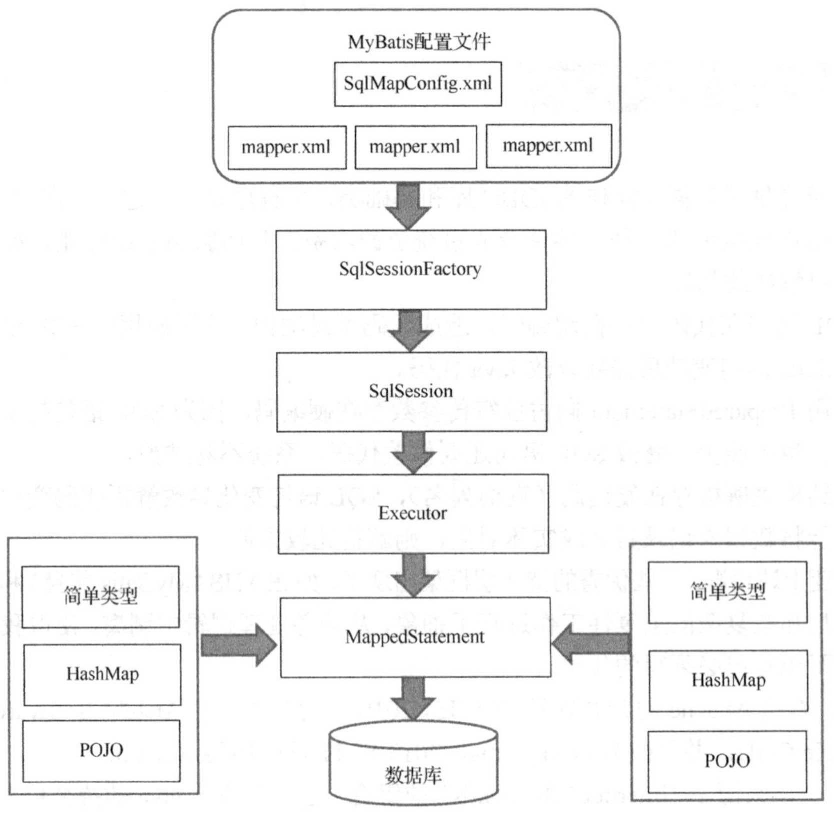

图 5-1　MyBatis 框架的架构

对于图 5-1 所示的架构图，我们来一一阐述每个环节：

（1）最顶部的 MyBatis 配置文件 SqlMapConfig.xml 作为 MyBatis 的全局配置文件，配置了 MyBatis 的运行环境等信息。mapper.xml 文件即 SQL 映射文件，其配置了操作数据库的 SQL 语句，此文件需要在 SqlMapConfig.xml 文件中进行加载。

（2）通过 MyBatis 的配置信息构造会话工厂 SqlSessionFactory，然后由会话工厂创建会话 SqlSession，通过 SqlSession 才能操作数据库。

（3）MyBatis 底层定义了使用 Executor（执行器）接口操作数据库。Executor 接口有两个实现，一个是基本执行器，另一个是缓存执行器。

（4）MappedStatement 也是 MyBatis 的一个底层封装对象，它包装了 MyBatis 配置信息及 SQL 映射信息等。mapper.xml 文件中每一个 SQL 语句对应一个 MappedStatement 对象，SQL 的唯一 ID 即 MappedStatement 对象的 ID。图 5-1 的左边表明 MappedStatement 对象可以对 SQL 的输入参数进行定义，包括简单类型、集合类型 HashMap、POJO 等。Executor 在执行 SQL 语句前会将输入的 Java 对象的属性值映射至该 SQL 语句中，此 SQL 语句是由 PreparedStatement 对象创建的 JDBC 预编译语句。图 5-1 的右边表明 MappedStatement 对象也可以对 SQL 语句的输出结果进行定义，包括简单类型、集合类型 HashMap、POJO。Executor 在执行 SQL 语句后通过该对象将输出结果映射至封装了 Java 对象的集合中，输出结果的映射过程相当于 JDBC 编程中对结果集的解析处理过程。

5.2 MyBatis Generator 工具

MyBatis 项目下有一个工具项目，即 MyBatis Generator，用来方便开发者生成项目数据库对应的 Model、Mapper、DAO 持久层代码。MyBatis Generator 提供了 Maven plugin、Ant target、Java 三种方式启动。现在主流的构建工具是 Maven 和 Gradle，虽然 MyBatis Generator 工具没有提供 Gradle 插件，但 Gradle 可以调用 Ant 任务，因此 Gradle 也能启动 MyBatis Generator。

5.2.1 使用 MyBatis Generator 工具前的数据库准备

下面我们继续沿用第 4 章的项目，演示如何使用 MyBatis Generator 工具，生成附录 A 中的 student.sql 数据库的持久层。

首先安装好 MySQL 数据库服务器和客户端工具软件 Navicat。

在配套教学资源视频里，我们使用的是 MySQL 5.6 和 Navicat 11。如果 MySQL 是绿色解压版的，请先在 CMD 下使用如下命令行注册并启动服务：

```
-->mysqld install 服务名
-->net start 服务名
```

进入 Navicat 工具，首先将 MySQL 服务器的 root 用户的密码设置为“123”。然后创建数据库 test，字符集设置为 UTF-8。再运行配套教学资源里的 student.sql 文件，创建表。创建成功后，test 数据库里一共有 5 张表：student（学生）、teacher（教师表）、course（课程表）、score（成绩表）、admin（管理员表）。表中具体的字段信息和表之间的关系请参考附录 A。

5.2.2　任务一：项目中自动生成 MyBatis 框架的持久层代码

以下是 MyBatis Generator 工具自动生成持久层代码的具体步骤，里面用到的文件在配套教学资源第 5 章的“5.2 自动生成与集成”里可以找到。

步骤 1．在 resources 下创建/mybatis/db-mysql.properties 文件，定义数据库的连接信息，每个属性的说明参看注释。具体内容如下：

```
# JDBC URL: jdbc:mysql:// + 数据库主机地址 + 端口号 + 数据库名
jdbc.url=jdbc:mysql://localhost:3306/test
# JDBC 用户名及密码
jdbc.user=root
jdbc.pass=123
# JDBC 驱动程序类
jdbc.driverClassName=com.mysql.jdbc.Driver
# 生成实体类所在的包
package.model=com.ssm.entity
# 生成 mapper 接口所在的包
package.mapper=com.ssm.dao
# 生成 mapper.xml 文件所在的包，默认存储在 resources 目录下
package.xml=mapper
```

步骤 2．在 resources 目录下创建/mybatis/generatorConfig.xml 文件，用于定义自动生成时的配置信息。内容如下：

```
<?xml version="1.0" encoding="UTF-8"?>
<!DOCTYPE generatorConfiguration
        PUBLIC "-//Mybatis.org//DTD Mybatis Generator Configuration 1.0//EN"
        "http://Mybatis.org/dtd/Mybatis-generator-config_1_0.dtd">
<generatorConfiguration>
    <context id="MysqlContext" targetRuntime="Mybatis3" defaultModelType="flat">
        <property name="beginningDelimiter" value="`"/>
        <property name="endingDelimiter" value="`"/>
        <commentGenerator>
            <property name="suppressDate" value="true"/>
        </commentGenerator>
        <jdbcConnection driverClass="${driverClass}"
                        connectionURL="${connectionURL}"
                        userId="${userId}"
                        password="${password}">
        </jdbcConnection>
        <javaModelGenerator targetPackage="${modelPackage}" targetProject="${src_
        main_java}"/>
        <sqlMapGenerator targetPackage="${sqlMapperPackage}" targetProject="${src_
        main_resources}"/>
        <javaClientGenerator targetPackage="${mapperPackage}" targetProject="${src
        _main_java}"
                             type="XMLMAPPER"/>
        <!-- sql 占位符，表示所有的表 -->
        <table tableName="%" enableCountByExample="false" enableUpdateByExample=
        "false"
```

```
            enableDeleteByExample="false" enableSelectByExample="false"
            selectByExampleQueryId="false">
        </table>
    </context>
</generatorConfiguration>
```

其中，前面的 context 标签的 targetRuntime 属性定义了运行的目标环境 MyBatis。defaultModelType 属性很重要，它定义了如何生成实体类。flat 模型将为每一张表只生成一个实体类。这个实体类包含表中的所有字段。这种模型最简单，推荐使用。另外还有 conditional、hierarchical 两种可选。如果表有主键，那么这两种模型会产生一个单独的主键实体类。如果表还有 BLOB 字段，则会为表生成一个包含所有 BLOB 字段的单独的实体类，然后为所有其他的字段生成一个单独的实体类。MyBatis Generator 会在所有生成的实体类之间维护一个继承关系。

jdbcConnection 标签定义了数据库连接属性。javaModelGenerator 标签定义了生成的实体类所存放的目录，sqlMapGenerator 标签定义了封装 SQL 语句的 mapper.xml 系列文件存放的目录，javaClientGenerator 标签定义了生成的数据访问对象 DAO 所存放的目录。它们共同使用的常量则来自前面定义的 jdbc.properties 文件，在下面的步骤 3 中会定义如何获取这些常量。

步骤 3. 修改 build.gradle 文件，添加自动生成工具 MyBatis Generator 的 Ant 任务。里面有 3 处变动，具体内容如下：

（1）添加配置。

```
configurations {
    MybatisGenerator
}
```

（2）添加 MybatisGenerator 需要的三个依赖：Mybatis-generator-core、mysql-connector-java、tk.Mybatis。

```
MybatisGenerator 'org.Mybatis.generator:Mybatis-generator-core:1.3.2'
MybatisGenerator 'mysql:mysql-connector-java:5.1.38'
MybatisGenerator 'tk.Mybatis:mapper:3.3.1'
```

（3）定义从 db-mysql.properties 属性文件读取属性常量。定义任务 MybatisGenerate，读取属性常量，读取 generatorConfig.xml 配置文件，执行类名为 GeneratorAntTask 的任务。

```
def getDbProperties = {
    def properties = new Properties()
    file("src/main/resources/Mybatis/db-mysql.properties").withInputStream
{ inputStream ->
        properties.load(inputStream)
    }
    properties;
}
task MybatisGenerate << {
    def properties = getDbProperties()
    ant.properties['targetProject'] = projectDir.path
    ant.properties['driverClass'] = properties.getProperty("jdbc.driverClassName")
    ant.properties['connectionURL'] = properties.getProperty("jdbc.url")
    ant.properties['userId'] = properties.getProperty("jdbc.user")
    ant.properties['password'] = properties.getProperty("jdbc.pass")
    ant.properties['src_main_java'] = sourceSets.main.java.srcDirs[0].path
```

```
    ant.properties['src_main_resources'] = sourceSets.main.resources.srcDirs[0].path
    ant.properties['modelPackage'] = properties.getProperty("package.model")
    ant.properties['mapperPackage'] = properties.getProperty("package.mapper")
    ant.properties['sqlMapperPackage'] =properties.getProperty("package.xml")
    ant.taskdef(
            name: 'mbgenerator',
            classname: 'org.Mybatis.generator.ant.GeneratorAntTask',
            classpath: configurations.MybatisGenerator.asPath
    )
    ant.mbgenerator(overwrite: true,
            configfile: 'src/main/resources/Mybatis/generatorConfig.xml', verbose:
true) {
        propertyset {
            propertyref(name: 'targetProject')
            propertyref(name: 'userId')
            propertyref(name: 'driverClass')
            propertyref(name: 'connectionURL')
            propertyref(name: 'password')
            propertyref(name: 'src_main_java')
            propertyref(name: 'src_main_resources')
            propertyref(name: 'modelPackage')
            propertyref(name: 'mapperPackage')
            propertyref(name: 'sqlMapperPackage')
        }
    }
}
```

步骤 4. 单击右侧边栏的 Gradle，先刷新 StudentGradle 模块，再展开 StudentGradle→Tasks→other→mybatisGenerate，如图 5-2 所示。

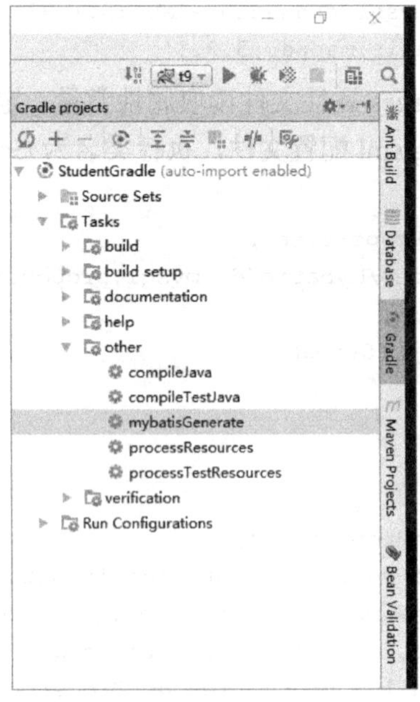

图 5-2 mybatisGenerate 任务所在位置

双击 mybatisGenerate 任务后，运行结果如图 5-3 所示。1 处控制台上显示 BUILD SUCCESSFUL 表示运行成功，2 处 Project 树状列表里展开的 com.ssm dao、com.ssm entity、mapper 目录是生成的数据库持久层文件的存放路径，里面有每张表所对应的实体类、mapper.xml 文件及 DAO 接口。

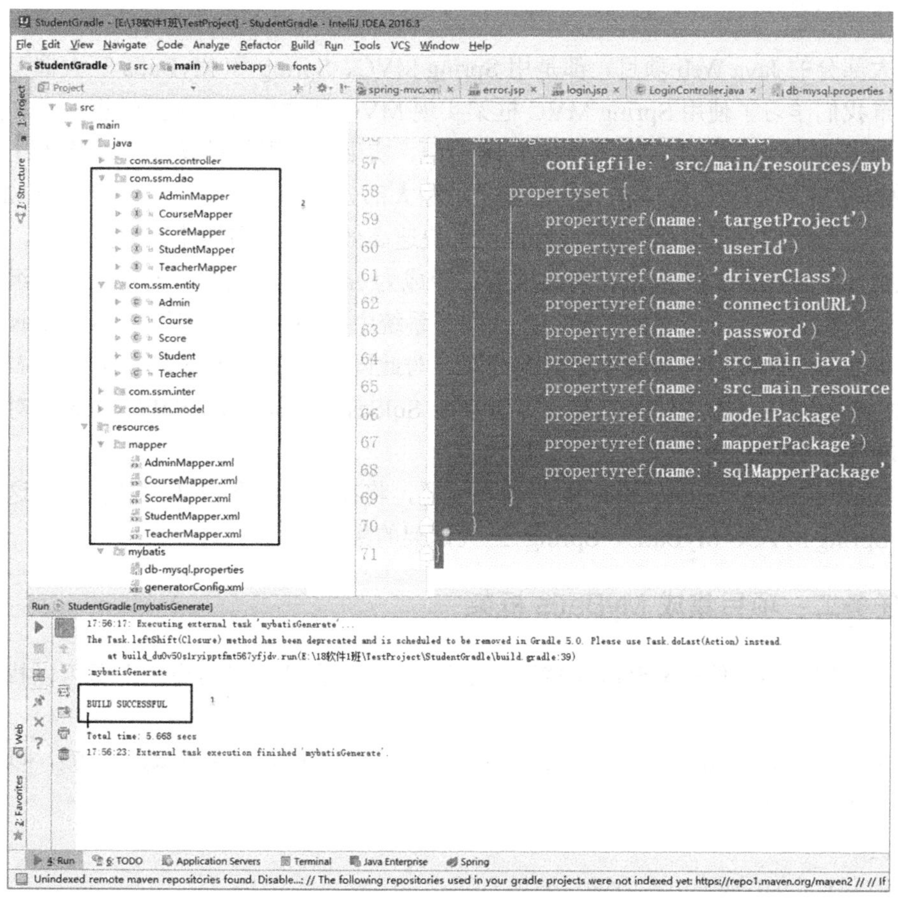

图 5-3　运行结果

下面教大家一个调试的小技巧。因为在运行任务的时候，难免出现各种错误。如出现图 5-4 所示的界面，就是 MyBatis Generator 运行有错的表现。单击红色箭头所指向的按钮，会显示具体的错误信息。根据错误提示修改代码后再运行。

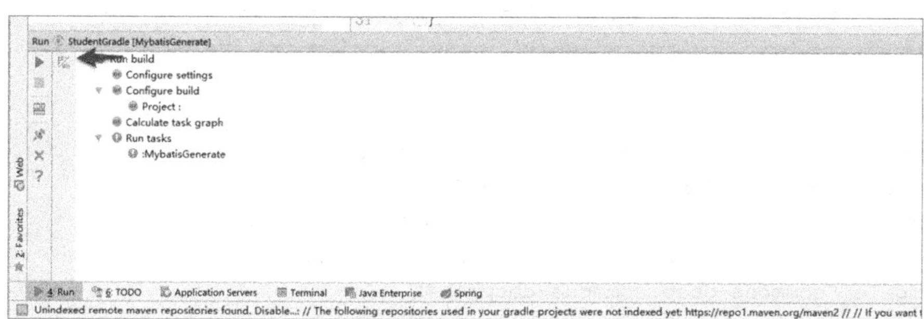

图 5-4　MyBatis Generator 运行有错

5.3 SSM 框架的总集成

5.3.1 集成简介

目前大部分的 Java Web 项目，都是用 Spring MVC、Spring 和 MyBatis 三大框架搭建而成的。第 4 章我们学习了使用 Spring MVC 框架实现 MVC 开发模式。后面的第 6 章、第 7 章将学习使用 Spring 的 IoC 有效地管理各种 Java 资源，实现即插即拔的功能，以及通过 Spring 的 AOP 将数据库事务委托给 Spring IoC 管理，消除大部分的事务代码，再配合本章 MyBatis 框架的高灵活、可配置、可优化 SQL 等特性，可快速高效地构建高性能的大型网站。

毫无疑问，MyBatis 和 Spring 两大框架已经成为 Java 互联网技术的主流框架组合，它们经受住了大数据和大批量请求的考验，在互联网系统中得到了广泛的应用。MyBatis 和 Spring 的集成使得业务层和模型层得到了更好的分离。与此同时，在 Spring 环境中使用 MyBatis 框架也更加简单，减少了不少代码，甚至可以不用 SqlSessionFactory、SqlSession 等对象，因为 Mybatis-Spring 为我们封装了它们。

关于 Spring IoC 和 Spring AOP 的原理和配置，将在第 6 章和第 7 章讲解。在这里，先教大家实现 Spring MVC、MyBatis、Spring 三大框架的集成。

5.3.2 任务二：项目集成 MyBatis 框架

下面所使用到的文件在配套教学资源第 5 章的“5.2 自动生成与集成”里可以找到。具体步骤如下：

步骤 1．在 build.gradle 文件中添加 MyBatis、Spring 和 MySQL 驱动的依赖，以及一些公共资源包。代码如下：

```
/Mybatis
    compile "org.Mybatis:Mybatis:3.4.1"
//Mybatis Spring 插件
  compile "org.Mybatis:Mybatis-Spring:1.3.1"
// https://mvnrepository.com/artifact/mysql/mysql-connector-java
    compile group: 'mysql', name: 'mysql-connector-java', version: '5.1.18'
// https://mvnrepository.com/artifact/org.Springframework/Spring-jdbc
compile group: 'org.Springframework', name: 'Spring-jdbc', version: '4.3.18. RELEASE'
// https://mvnrepository.com/artifact/org.Springframework/Spring-tx
compile group: 'org.Springframework', name: 'Spring-tx', version: '4.3.18.RELEASE'

//公共资源包
    compile "commons-logging:commons-logging:1.2"
    compile "commons-lang:commons-lang:2.6"
    compile "org.apache.commons:commons-collections4:4.0"
    compile "commons-beanutils:commons-beanutils:1.8.3"
    compile "commons-dbcp:commons-dbcp:1.4"
    compile "commons-pool:commons-pool:1.6"
```

步骤 2．在 resources 文件夹下创建数据库连接的属性文件 jdbc.properties，定义数据库连接需要的信息和参数，如驱动、连接字符串、用户名和密码等。代码如下：

```
jdbc.driver=com.mysql.jdbc.Driver
jdbc.url=jdbc:mysql://localhost:3306/test
jdbc.username=root
jdbc.password=123
#定义初始连接数
initialSize=0
#定义最大连接数
maxActive=20
#定义最大空闲
maxIdle=20
#定义最小空闲
minIdle=1
#定义最长等待时间 1 分钟
maxWait=60000
```

步骤 3. 在 resources 文件夹下添加 Spring 集成 MyBatis 的配置文件 spring-mybatis.xml，使用 Apache 的数据库连接池 BasicDataSource 创建数据源 dataSource，在子标签里定义其使用的连接信息。再将 dataSource 注入 SqlSessionFactoryBean 工厂 sqlSessionFactory，定义其扫描的 mapper.xml 文件位于类路径的 resources/mapper 下。后面定义了映射扫描配置器 MapperScannerConfigurer，用于自动扫描 DAO 接口所在包 com.ssm.dao。代码如下：

```
<?xml version="1.0" encoding="UTF-8"?>
<beans xmlns="http://www.Springframework.org/schema/beans"
      xmlns:xsi="http://www.w3.org/2001/XMLSchema-instance"
      xmlns:context="http://www.Springframework.org/schema/context"
      xmlns:tx="http://www.Springframework.org/schema/tx"
      xsi:schemaLocation="http://www.Springframework.org/schema/beans
      http://www.Springframework.org/schema/beans/Spring-beans.xsd
      http://www.Springframework.org/schema/context
      http://www.Springframework.org/schema/context/Spring-context.xsd
      http://www.Springframework.org/schema/tx
      http://www.Springframework.org/schema/tx/Spring-tx.xsd">
    <!-- 配置数据库相关参数 properties 的属性: ${url} -->
    <context:property-placeholder location="classpath*:jdbc.properties"/>
    <!-- 数据库连接池 -->
    <bean id="dataSource" class="org.apache.commons.dbcp.BasicDataSource" destroy-
    method="close">
        <property name="driverClassName" value="${jdbc.driver}"/>
        <property name="url" value="${jdbc.url}"/>
        <property name="username" value="${jdbc.username}"/>
        <property name="password" value="${jdbc.password}"/>
    </bean>
    <!-- Spring 和 MyBatis 完美整合，不需要 MyBatis 的配置映射文件 -->
    <bean id="sqlSessionFactory" class="org.Mybatis.Spring.SqlSession FactoryBean">
        <property name="dataSource" ref="dataSource" />
        <!-- 自动扫描 mapping.xml 文件 -->
        <property name="mapperLocations" value="classpath*:/mapper/*.xml"> </property>
    </bean>
```

```
    <!-- DAO 接口所在包名，Spring 会自动查找其下的类 -->
    <bean class="org.Mybatis.Spring.mapper.MapperScannerConfigurer">
        <property name="basePackage" value="com.ssm.dao" />
        <property name="sqlSessionFactoryBeanName" value="sqlSessionFactory"> </property>
    </bean>
    <bean id="sqlSessionTemplate" class="org.Mybatis.Spring.SqlSessionTemplate">
        <constructor-arg index="0" ref="sqlSessionFactory" />
    </bean>
    <!-- (事务管理)transaction manager, use JtaTransactionManager for global tx -->
    <bean id="transactionManager"
          class="org.Springframework.jdbc.datasource.DataSourceTransactionManager">
        <property name="dataSource" ref="dataSource" />
    </bean>
</beans>
```

步骤 4. 修改 web.xml 文件，添加 Spring 的上下文加载监听器 ContextLoaderListener，它启动时会加载上面配置的 spring-mybatis.xml 文件。代码如下：

```
<!-- 配置 Spring 的监听器 -->
<!--配置启动 IoC 容器的 Listener-->
<!-- 设置 Spring 容器加载配置文件的路径 -->
<context-param>
   <param-name>contextConfigLocation</param-name>
   <param-value>classpath:spring-mybatis.xml</param-value>
</context-param>
<listener>  <listener-class>org.Springframework.web.context.Context LoaderListener
</listener-class>
</listener>
```

到此为止，三大框架的集成完成了。部署上面的项目，启动 Tomcat 测试，如果控制台没有报错，首页可以访问，则集成成功。

5.4 mapper.xml 文件的编写

在 5.2 节，resources/mapper 下生成了每个数据库表对应的 mapper.xml 文件，如 student 表对应的就是 StudentMapper.xml。我们就以 StudentMapper.xml 为例，在文件注释里逐一对照，讲解其中标签和属性的意义。StudentMapper.xml 文件代码如下：

```
<?xml version="1.0" encoding="UTF-8" ?>
<!DOCTYPE mapper PUBLIC "-//Mybatis.org//DTD Mapper 3.0//EN" "http://Mybatis.org/
dtd/Mybatis-3-mapper.dtd">
<!-- ==================代理方式==================
由 mapper 标签开始，到/mapper 标签结束，定义了一个命名空间 namespace，在代理中使用。这个
namespace 是唯一的。 -->
<mapper namespace="com.ssm.dao.StudentMapper">
   <!-- =============映射关系标签=============
属性 id：查询结果集 resultMap 的唯一标识
属性 type：POJO 类的包名、类名
子标签 id：表 student 的主键 Id,对应属性 id，类型是整型 Integer
```

```
    子标签 result 的属性 column：查询出来的列名
  子标签 result 的属性 property：是 POJO 类里所指定的列名
  属性 jdbcType：定义列在数据库中的数据类型
      -->
    <resultMap id="BaseResultMap" type="com.ssm.entity.Student" >
    <!--
      WARNING - @mbggenerated
      This element is automatically generated by Mybatis Generator, do not modify.
    -->
    <id column="Id" property="id" jdbcType="INTEGER" />
    <result column="name" property="name" jdbcType="VARCHAR" />
    <result column="password" property="password" jdbcType="VARCHAR" />
    <result column="sex" property="sex" jdbcType="INTEGER" />
    <result column="clazz" property="clazz" jdbcType="VARCHAR" />
    <result column="birthday" property="birthday" jdbcType="VARCHAR" />
    <result column="image" property="image" jdbcType="VARCHAR" />
   </resultMap>
  <!-- ==================定义 sql 片段==============
    sql:id 是 sql 片段标签属性，Base_Column_List 是该片段的唯一标识,定义它是所有列的集合。列
是可选的。 -->
     <sql id="Base_Column_List" >
    <!--
      WARNING - @mbggenerated
      This element is automatically generated by Mybatis Generator, do not modify.
    -->
    Id, name, password, sex, clazz, birthday, image
   </sql>
    <!-- 查找 select 标签里的 id：务必和接口里对应的方法名一致,
        resultMap：输出类型里写映射标签里的 id
        parameterType：输入类型，规范输入数据的类型，指明查询时使用的参数类型-->
      <!-- 通过主键查询修改用户信息的 id -->
   <select id="selectByPrimaryKey" resultMap="BaseResultMap" parameterType= "java.
lang.Integer" >
    <!--
      WARNING - @mbggenerated
      This element is automatically generated by Mybatis Generator, do not modify.
 -->
  <!-- 定义通过参数 id 查找 student 表的 SQL 语句 -->
    select
    <include refid="Base_Column_List" />
    from student
    where Id = #{id,jdbcType=INTEGER}
   </select>
    </mapper>
```

在 mapper.xml 文件中，<select>、<insert>、<update>、<delete> 分别对应查询、添加、更新、删除的 SQL 操作，参数可以使用正则表达式。

5.4.1 小知识：控制台跟踪数据库操作执行

在进行数据库操作前，教大家一个使用 log4j 在控制台跟踪数据库操作执行情况的方法，以便在后续的操作中能在控制台及时跟踪到出错信息。操作步骤如下：

步骤 1．在 build.gradle 中添加 log4j 包的依赖，内容如下：

```
// https://mvnrepository.com/artifact/log4j/log4j
compile group: 'log4j', name: 'log4j', version: '1.2.17'
```

步骤 2．在 resources 文件夹下新建 log4j.properties 属性文件，内容如下：

```
#log4j.rootLogger=CONSOLE,info,error,DEBUG
#调试 log4j 的错误时打开
log4j.rootLogger=info,error,CONSOLE,DEBUG
#上线时打开，打印错误
#log4j.rootLogger=error
log4j.appender.CONSOLE=org.apache.log4j.ConsoleAppender
log4j.appender.CONSOLE.layout=org.apache.log4j.PatternLayout
log4j.appender.CONSOLE.layout.ConversionPattern=%d{yyyy-MM-dd-HH-mm}  [%t]  [%c]
[%p] - %m%n

log4j.logger.info=info
log4j.appender.info=org.apache.log4j.DailyRollingFileAppender
log4j.appender.info.layout=org.apache.log4j.PatternLayout
log4j.appender.info.layout.ConversionPattern=%d{yyyy-MM-dd-HH-mm} [%t] [%c] [%p]
- %m%n
log4j.appender.info.datePattern='.'yyyy-MM-dd
log4j.appender.info.Threshold = info
log4j.appender.info.append=true
#log4j.appender.info.File=/usr/local/api/logs/info/info.log
#log4j.appender.info.File=/Users/yelongyang/Downloads/logs/info/info.log

log4j.logger.error=error
log4j.appender.error=org.apache.log4j.DailyRollingFileAppender
log4j.appender.error.layout=org.apache.log4j.PatternLayout
log4j.appender.error.layout.ConversionPattern=%d{yyyy-MM-dd-HH-mm} [%t] [%c] [%p]
- %m%n
log4j.appender.error.datePattern='.'yyyy-MM-dd
log4j.appender.console.Threshold=DEBUG

log4j.appender.error.append=true
#log4j.appender.error.File=/usr/local/api/logs/error/error.log
#log4j.appender.error.File=/Users/yelongyang/Downloads/logs/error/error.log

log4j.logger.DEBUG=DEBUG
log4j.appender.DEBUG=org.apache.log4j.DailyRollingFileAppender
log4j.appender.DEBUG.layout=org.apache.log4j.PatternLayout
log4j.appender.DEBUG.layout.ConversionPattern=%d{yyyy-MM-dd-HH-mm} [%t] [%c] [%p]
- %m%n
```

```
log4j.appender.DEBUG.datePattern='.'yyyy-MM-dd
log4j.appender.DEBUG.Threshold = DEBUG
log4j.appender.DEBUG.append=true
#log4j.appender.DEBUG.File=/usr/local/api/logs/debug/debug.log
#log4j.appender.DEBUG.File=/Users/yelongyang/Downloads/logs/debug/debug.log

### Debug，修改成项目 groupID
log4j.logger.com.ssm=DEBUG
log4j.logger.com.Mybatis.common.jdbc.SimpleDataSource=DEBUG
log4j.logger.com.Mybatis.common.jdbc.ScriptRunner=DEBUG
log4j.logger.com.Mybatis.sqlmap.engine.impl.SqlMapClientDelegate=DEBUG
log4j.logger.java.sql.Connection=DEBUG
log4j.logger.java.sql.Statement=DEBUG
log4j.logger.java.sql.PreparedStatement=DEBUG
log4j.logger.java.sql.ResultSet=DEBUG

##本地环境
log4j.appender.info.File=E:/xjl/logs/info/info.log
log4j.appender.error.File=E:/xjl/logs/error/error.log
log4j.appender.DEBUG.File=E:/xjl/logs/debug/debug.log

#生产环境
#log4j.appender.info.File=/usr/local/Spring-boot-api/logs/info/info.log
#log4j.appender.error.File=/usr/local/Spring-boot-api/logs/error/error.log
#log4j.appender.DEBUG.File=/usr/local/Spring-boot-api/logs/debug/debug.log
```

提示：在 log4j.properties 文件中，log4j.logger.com.ssm=DEBUG 这个语句需要特别注意，com.ssm 指的是自己项目的 groupID。

经过上面两个步骤后，执行后续的数据库访问时，就可以在控制台跟踪到数据库操作执行的信息。

5.4.2　任务三：显示所有学生信息功能的实现

【示例 5-1】功能：显示 student 表中所有学生的信息。

首先要搭建好项目的 Java 源码架构层，DAO 目录已经自动生成，新建 service 目录，controller 目录也已存在。代码如下：

步骤 1．在 resources/mapper/StudentMapper.xml 文件里，增添查询所有学生信息的 SQL 语句：

```
<select id="list" resultMap="BaseResultMap" >
    select
   <include refid="Base_Column_List" />
   from student
</select>
```

步骤 2．在 java/com.ssm.dao/StudentMapper.java 源文件里，增加对应 SQL 语句的方法的声明，方法名 list 要与步骤 1 里 select id 一致。代码如下：

```
List<com.ssm.entity.Student> list();
```

步骤 3．在 java 文件夹下新建包 com.ssm.service，在其下新建业务接口 StudentService，声明 list 业务方法。代码如下：

```
public interface StudentService {
    public List<Student> list();
}
```

步骤 4．在包 com.ssm.service 下新建子包 impl，新建实现业务类 StudentServiceImpl。代码如下：

```
@Service("studentServiceImpl")
public class StudentServiceImpl implements StudentService {
    @Autowired
    private StudentMapper studentMapper;
    @Override
    public List<Student> list() {
        return studentMapper.list();
    }
}
```

@Service 注解定义了一个由 StudentServiceImpl 类生成的 Service 类型的对象，名称是 studentServiceImpl，后面会在 StudentController 中用到。@Autowired（自动注入）注解，会把 Spring 容器的 studentMapper 对象注入进来。在后面第 6 章我们会讲到 Spring 容器如何创建并管理这些对象。

步骤 5．修改 java/com.ssm.controller/StudentController.java 源文件，添加 studentServiceImpl 对象的自动注入代码。为 List 方法增加 Model 类型的参数 model，该参数用于传递 students 对象。代码如下：

```
    @Autowired
     private StudentService studentServiceImpl;
    @RequestMapping("list.do")
    public String list(Model model){
    //调用 studentServiceImpl 的 list 方法，从数据库取数据
     List<Student> students=studentServiceImpl.list();
     //通过 model 对象带出 students 数据到 listStudent.jsp 页面
     model.addAttribute("students",students);
     return "listStudents";
}
```

步骤 6．修改 listStudents.jsp 文件，使用 Bootstrap 框架的条纹表格显示数据，其中还用到 jstl 标签中的<c:forEach> 标签和表达式的迭代显示。代码如下：

```
<table class="table table-striped">
    <tr>
        <td>id</td>
        <td>名字</td>
        <td>性别</td>
        <td>班级</td>
        <td>生日</td>
```

```
        <td>头像</td>
          </tr>
    <c:forEach items="${students}" var="student">
        <tr>
            <td>${student.id}</td>
            <td>${student.name}</td>
            <td>${student.sex}</td>
            <td>${student.clazz}</td>
            <td>${student.birthday}</td>
            <td>${student.image}</td>
        </tr>
    </c:forEach>
</table>
```

步骤 7. 部署运行，在浏览器地址栏里输入 http://localhost:8077/Student/list.do，运行结果如图 5-5 所示。

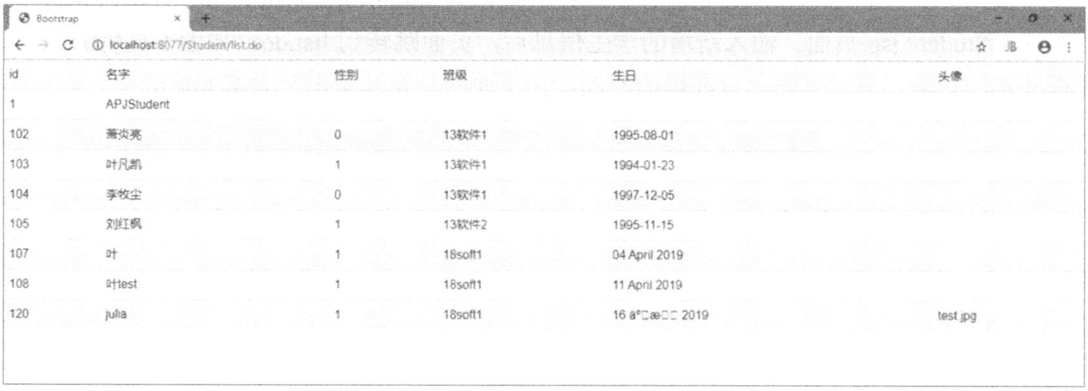

id	名字	性别	班级	生日	头像
1	APJStudent				
102	萧炎亮	0	13软件1	1995-08-01	
103	叶凡凯	1	13软件1	1994-01-23	
104	李牧尘	0	13软件1	1997-12-05	
105	刘红枫	1	13软件2	1995-11-15	
107	叶	1	18soft1	04 April 2019	
108	叶test	1	18soft1	11 April 2019	
120	julia	1	18soft1	16 å°□æ□□ 2019	test.jpg

图 5-5　示例 5-1 运行结果

5.4.3　任务四：增加学生功能的实现

【示例 5-2】功能：在 addStudent.jsp 输入学生的简单信息，并将学生信息添加到数据库 student 表中。具体代码和操作步骤如下：

提示：由于在 StudentMapper.xml 和 StudentMapper.java 文件中已经有增加学生的 SQL 语句和方法 insertSelective，所以不必修改，直接调用即可。

步骤 1. 修改业务实现接口 StudentService.java 源文件，增加 insert 方法的声明。代码如下：

```
public int insert(Student student);
```

步骤 2. 修改业务实现类 StudentServiceImpl.java 源文件，实现 insert 方法，里面调用了 studentMapper 的 insertSelective 方法，往 student 表里添加一条记录。代码如下：

```
@Override
   public int insert(Student student) {
       return studentMapper.insertSelective(student);   }
```

步骤 3. 对 StudentController.java 源文件的 add 方法做如下修改：调用 studentServiceImp 的 insert 方法后，重定向到 list 方法，显示新添加的学生信息。具体修改位置见源码：

```
          uploadFile(file);
        student.setImage(file.getOriginalFilename());
        // 添加记录
        studentServiceImpl.insert(student);
        return "redirect:list.do";
```

方法重定向到另一个方法时，在 return 里使用重定向 redirect:。

步骤 4．修改 addStudent.jsp 文件，丰富其他属性信息，性别使用 radio 控件，birthday 使用 datetimepicker 控件，我们在第 3 章演示过。代码如下：

```
  密码：<input name="password"  class="form-control" type="password" value= "${student.
pwd}">
  性别：<input name="sex"  class="form-control" type="text" value="${student. sex}">
  班级：<input name="clazz"  class="form-control" type="text" value="${student. clazz}">
  生日：<input name="birthday"  class="form-control" type="text" value="${student.
birthday}">
```

步骤 5．部署运行，打开浏览器，在地址栏里输入 http://localhost:8077/Student/toAdd.do，进入 AddStudent.jsp 页面。输入新增的学生信息后，页面跳转到 list.do，如图 5-6 所示，显示了新添加的记录。关于页面上有乱码的问题，在下面的补充知识里教大家如何改正。

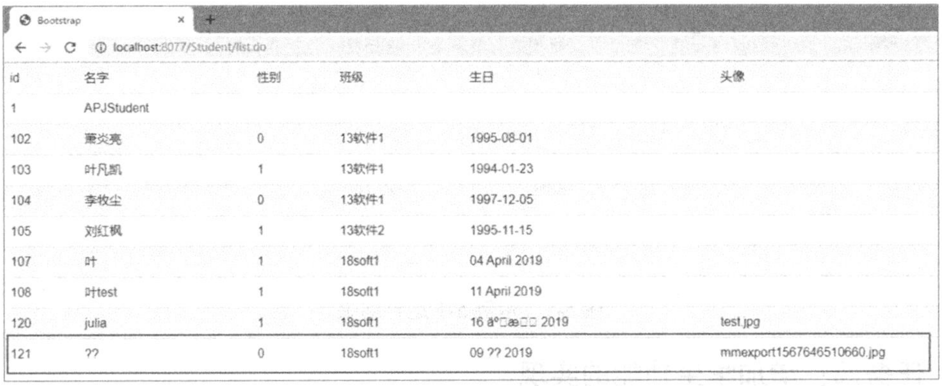

Bootstrap

localhost:8077/Student/list.do

id	名字	性别	班级	生日	头像
1	APJStudent				
102	萧炎亮	0	13软件1	1995-08-01	
103	叶凡凯	1	13软件1	1994-01-23	
104	李牧尘	0	13软件1	1997-12-05	
105	刘红枫	1	13软件2	1995-11-15	
107	叶	1	18soft1	04 April 2019	
108	叶test	1	18soft1	11 April 2019	
120	julia	1	18soft1	16 å°□æ□□ 2019	test.jpg
121	??	0	18soft1	09 ?? 2019	mmexport1567646510660.jpg

图 5-6　增加新学生信息

5.4.4　补充知识：解决中文乱码问题

除了在 MySQL 数据库创建的时候选择 UTF-8 字符集项目，还有几个关键操作步骤，用于解决中文乱码问题。

（1）在所有显示中文的 JSP 页面上添加下面这段指令：

```
  %@page  pageEncoding="UTF-8"  contentType="text/html; utf-8" %
```

（2）修改 resources 目录下的 jdbc.properties 文件，在 jdbc.url 属性后添加?useUnicode=true&characterEncoding=UTF-8，表示数据库使用 Unicode 字符集。代码如下：

```
  jdbc.url=jdbc:mysql://localhost:3306/test?useUnicode=true&characterEncoding=UT
F-8
```

（3）修改 webapp/WEB-INF 下的 web.xml 文件，添加字符编码过滤器。代码如下：

```
<!--解决 POST 乱码问题-->
<filter>
    <filter-name>CharacterEncodingFilter</filter-name>
    <filter-class>org.Springframework.web.filter.CharacterEncodingFilter</filter-
    class>
    <init-param>
        <param-name>encoding</param-name>
        <param-value>utf-8</param-value>
    </init-param>
</filter>
<filter-mapping>
    <filter-name>CharacterEncodingFilter</filter-name>
    <url-pattern>/*</url-pattern>
</filter-mapping>
```

乱码问题解决后的运行效果如图 5-7 所示。

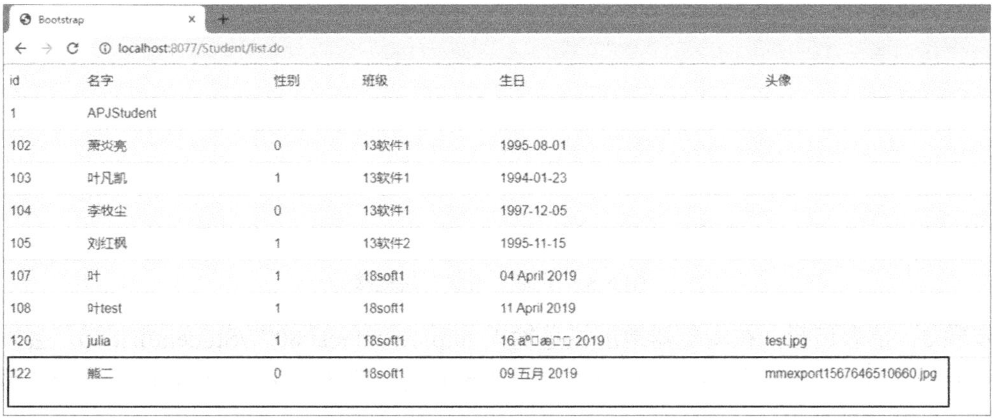

Bootstrap

localhost:8077/Student/list.do

id	名字	性别	班级	生日	头像
1	APJStudent				
102	萧炎亮	0	13软件1	1995-08-01	
103	叶凡凯	1	13软件1	1994-01-23	
104	李牧尘	0	13软件1	1997-12-05	
105	刘红枫	1	13软件2	1995-11-15	
107	叶	1	18soft1	04 April 2019	
108	叶test	1	18soft1	11 April 2019	
120	julia	1	18soft1	16 aº□æ□□ 2019	test.jpg
122	熊二	0	18soft1	09 五月 2019	mmexport1567646510660.jpg

图 5-7　解决乱码后的运行效果

5.4.5　任务五：删除学生功能的实现

【示例 5-3】功能：在 listStudents.jsp 页面增加“删除”超链接，单击该超链接后完成删除，并显示新的数据。

由于在 StudentMapper.xml 和 StudentMapper.java 文件中已经有删除学生的 SQL 语句和方法 deleteByPrimaryKey，所以不必修改，直接调用即可。其他操作步骤和代码如下：

步骤 1. 修改业务接口 StudentService.java 源文件，添加删除方法 delete 的声明，通过入参主键 id 删除。代码如下：

```
public int delete(int id);
```

步骤 2. 修改业务实现类 StudentServiceImpl.java 源文件，调用 studentMapper 对象的 delete ByPrimaryKey 方法实现删除操作。代码如下：

```
@Override
    public int delete(int id) {
```

```
        return studentMapper.deleteByPrimaryKey(id);
    }
```

步骤 3．修改 StudentController.java 源文件，添加删除方法；通过 delete.do 请求映射该方法，方法有一个入参 id，方法体内调用业务实现类的 delete 方法，删除完毕后重定向到 list.do。代码如下：

```
  @RequestMapping("delete.do")
      public String delete(int id){
          studentServiceImpl.delete(id);
          return "redirect:list.do";
    }
```

步骤 4．修改 listStudents.jsp 文件，给每条记录增加“删除”超链接。代码如下：

```
<td><a href="delete.do?id=${student.id}">删除</a></td>
```

增加“删除”超链接效果如图 5-8 所示。

id	名字	性别	班级	生日	头像	编辑
102	萧炎亮	0	13软件1	1995-08-01		删除
103	叶凡凯	1	13软件1	1994-01-23		删除
104	李牧尘	0	13软件1	1997-12-05		删除
105	刘红枫	1	13软件2	1995-11-15		删除
107	叶	1	18soft1	04 April 2019		删除
108	叶test	1	18soft1	11 April 2019		删除
120	julia	1	18soft1	16 a°□æ□□ 2019	test.jpg	删除
122	熊二	0	18soft1	09 五月 2019	mmexport1567646510660.jpg	删除

图 5-8　增加“删除”超链接

步骤 5．部署运行，在浏览器地址栏里输入 http://localhost:8077/Student/list.do，会发现每条记录后面都有一个“删除”超链接，单击该超链接则会删除该条记录，然后自动刷新该页面，显示删除后的其他数据。

5.4.6　任务六：修改学生信息功能的实现

【示例 5-4】在示例 5-3 基础上为 listStudents.jsp 增加“修改”超链接，单击该超链接后进入 updateStudent.jsp 页面，修改提交后，重新回到 listStudents.jsp 页面。

功能分析：修改动作实际上是两个步骤，首先单击“修改”超链接，通过主键 id 找到需要修改的学生信息，跳转到 updateStudent.jsp 页面，用户修改学生信息后，提交到数据库，然后回到 listStudents.jsp 页面中显示。

由于 StudentMapper.xml 和 StudentMapper.java 文件中已经有修改学生的 SQL 语句和方法 updateByPrimaryKeySelective，所以不必修改，直接调用即可。下面是具体的操作步骤和代码：

步骤 1．修改业务接口 StudentService.java 源文件，添加通过主键查找和修改学生信息的方法声明。代码如下：

```
      public Student find(int id);
      public int update(Student student);
```

步骤 2．修改业务实现类 StudentServiceImpl.java 源文件，实现通过主键查找和修改学生信息的两个方法。代码如下：

```
    @Override
    public Student find(int id) {
        return studentMapper.selectByPrimaryKey(id);
    }
    @Override
    public int update(Student student) {
        return studentMapper.updateByPrimaryKey(student);
}
```

步骤 3．修改 StudentController.java 源文件，添加 toUpdate 和 update 方法，toUpdate 方法通过主键找到需要修改的学生信息，通过 model 封装数据 student 后，跳转到 updateStudent.jsp 页面。用户修改完，提交修改，跳转到 update 方法，调用 studentServiceImpl 的 update 方法实现对数据库的更新操作，最后跳转到 listStudents.jsp 页面，显示学生信息。代码如下：

```
@RequestMapping("toUpdate.do")
    public String toUpdate(int id,Model model){
        Student student=studentServiceImpl.find(id);
        model.addAttribute("student",student);
        return "updateStudent";
    }
    @RequestMapping("update.do")
    public String update(Student student){
        studentServiceImpl.update(student);
        return "redirect:list.do";
}
```

步骤 4．参照示例 5-3，修改 listStudents.jsp 文件，添加“修改”超链接。代码如下：

```
<a href="toUpdate.do?id=${student.id}">修改</a>
```

步骤 5．在 webapp/WEB-INF/views 下，新建修改学生信息的文件 updateStudent.jsp，通过 value=${student.*}的方式来接收数据库传递过来的学生信息。代码如下：

```
<%@page pageEncoding="UTF-8" %>
<%@taglib prefix="c" uri="http://java.sun.com/jsp/jstl/core" %>
<%@ taglib prefix="form" uri="http://www.Springframework.org/tags/form" %>
<html>
<head>
    <meta charset="utf-8">
    <title>Bootstrap </title>
    <link href="/bootstrap/css/bootstrap.min.css" rel="stylesheet">
    <link href="/bootstrap/css/bootstrap-datetimepicker.min.css" rel="stylesheet">
</head>
<body>
<!-- jQuery 文件。务必在 bootstrap.min.js 之前引入 -->
<script src="/bootstrap/js/jquery-3.3.1.min.js"></script>
<!-- 最新的 Bootstrap 核心 JavaScript 文件 -->
<script src="/bootstrap/js/bootstrap.min.js"></script>
<script src="/bootstrap/js/bootstrap-datetimepicker.min.js"></script>
```

```
<form:form  action="update.do"  method="post"  class="form-control"  enctype=
"multipart/form-data">
    姓名：<input name="name"  class="form-control" type="text" value="${student.
name}">
    密码：<input name="password"  class="form-control" type="password" value=
"${student.password}">
    性别：<div>
        <label class="radio-inline">
            <input  type="radio"  name="sex"  id="optionsRadios3"  value="0"  <c:if
test="${student.sex eq '0'}">checked</c:if>/> 男
        </label>
        <label class="radio-inline">
            <input  type="radio"  name="sex"  id="optionsRadios4"  value="1"  <c:if
test="${student.sex eq '1'}">checked</c:if>/> 女
        </label>
    </div>
    班级：<input name="clazz"  class="form-control" type="text" value="${student.
clazz}">
    <div class="form-group">
        <label for="dtp_input2" class="col-md-2 control-label">生日</label>
        <div  class="input-group  date  form_date  col-md-5"  data-date=""
data-date-format="dd MM yyyy" data-link-field="dtp_input2" data-link-format= "yyyy-
mm-dd">
            <input class="form-control" size="16" type="text" name="birthday" value=
"${student.birthday}" readonly>
            <span  class="input-group-addon"><span  class="glyphicon  glyphicon-
remove"></span></span>
            <span  class="input-group-addon"><span  class="glyphicon  glyphicon-
calendar"></span></span>
        </div>
        <input type="hidden" id="dtp_input2" value="" /><br/>
    </div>
    <div class="form-group">
        <label for="file" class="col-sm-2 control-label">头像：</label>
        <div class="form-group">
            <input type="file" class="form-control" id="file" name="file" placeholder
="head image" value="${student.image}">
        </div>
    </div><br/>
    <div class="form-group">
    <input  class="col-lg-4" value="update student" type="submit">
    </div>
</form:form>
</body>
<script src="../bootstrap/js/locales/bootstrap-datetimepicker.zh-CN.js"></script>
<script type="text/javascript">
    $('.form_date').datetimepicker({
        language:  'zh-CN',
```

```
        weekStart: 1,
        todayBtn:  1,
        autoclose: 1,
        todayHighlight: 1,
        startView: 2,
        minView: 2,
        forceParse: 0
    });
</script>
</html>
```

该文件中有两个地方要注意。第一个是使用了 radio 控件来实现性别的单选，第二个是使用了第 3 章学过的 datetimepicker 控件来选择生日。

步骤 6．部署运行，在浏览器地址栏里输入 http://localhost:8077/Student/list.do，如图 5-9 所示。准备修改 id 为“120”的记录的名字 julia，单击“修改”超链接。

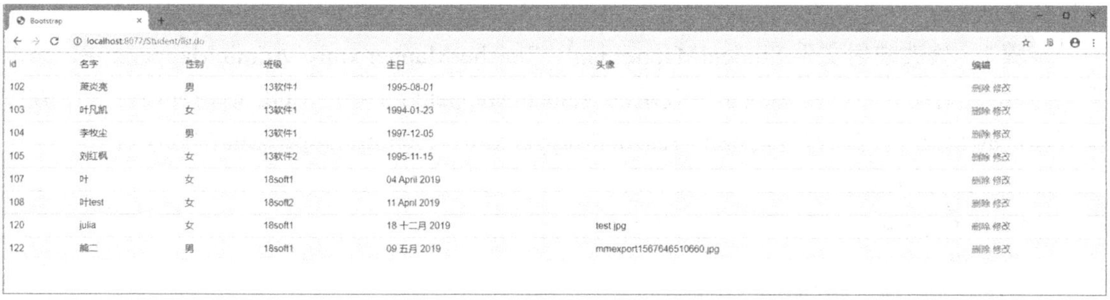

图 5-9　增加“修改”超链接

执行上述操作后会跳转到如图 5-10 所示的 updateStudent.jsp 页面，修改姓名 julia 为 athena。单击 update student 按钮，再次重定向到 listStudents.jsp 页面，显示修改后的数据内容，如图 5-11 所示。

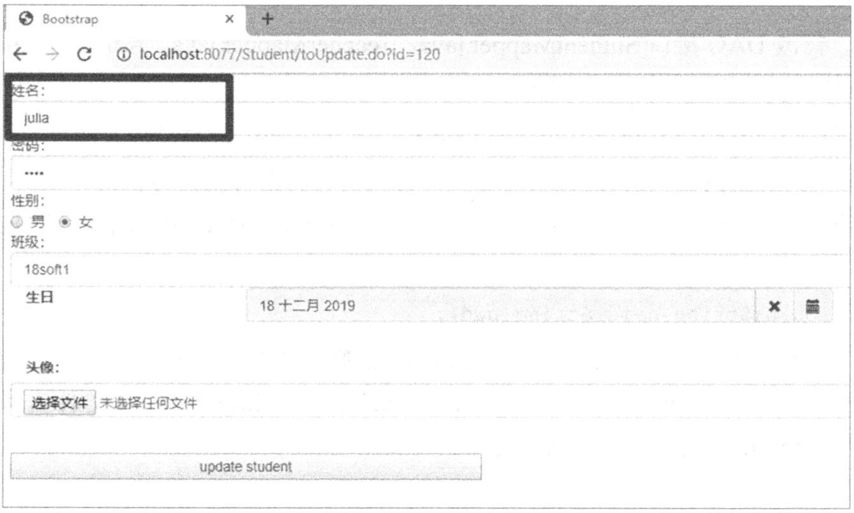

图 5-10　在 updateStudent.jsp 页面修改姓名

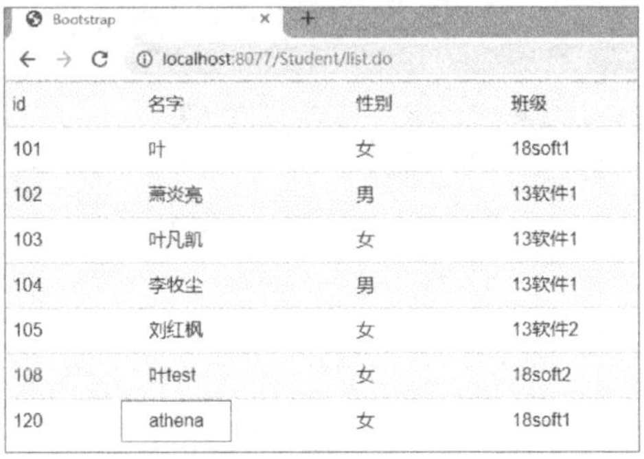
Bootstrap
localhost:8077/Student/list.do
id 名字 性别 班级
101 叶 女 18soft1
102 萧炎亮 男 13软件1
103 叶凡凯 女 13软件1
104 李牧尘 男 13软件1
105 刘红枫 女 13软件2
108 叶test 女 18soft2
120 athena 女 18soft1

图 5-11　更新后数据显示效果

5.4.7　拓展任务：学生登录功能的实现

【示例 5-5】实现 login.jsp 页面的登录功能，并设置用户名为 Session 类型变量，实现登录的拦截功能。

步骤 1．修改映射文件 StudentMapper.xml、TeacherMapper.xml、AdminMapper.xml，添加登录的 SQL 语句。下面是实现学生登录的 StudentMapper.xml 中的 SQL 语句片段，关于教师和管理员的 SQL 语句请参看对应的 TeacherMapper.xml、AdminMapper.xml 源码文件。

```
 <select id="login"
resultMap="BaseResultMap"  parameterType="java.lang.String" >
    select
    <include refid="Base_Column_List" />
    from student
where name=#{name,jdbcType=VARCHAR}
and password=#{pwd,jdbcType=VARCHAR}
  </select>
```

上面的 SQL 语句定义了两个入参 name 和 password，分别代表学生登录时输入的用户名和密码。

步骤 2．修改 DAO 接口 StudentMapper.java、TeacherMapper.java、AdminMapper.java 源文件，分别添加 login 方法的声明。代码如下：

```
Student login(@Param("name") String name,@Param("pwd") String pwd);
```

步骤 3．修改业务接口 StudentService.java 源文件，添加 login 方法的声明。另外，新建 TeacherService.java、AdminService.java 文件，具体内容与 StudentService.java 文件类似。代码如下：

```
Student login(String name,String pwd);
```

步骤 4．修改业务实现类 StudentServiceImpl.java 源码，实现 login 方法，调用 DAO 的 login 方法。另外，新建 TeacherServiceImpl.java、AdminServiceImpl.java 源文件，具体内容与 StudentServiceImpl.java 文件类似。代码如下：

```
     @Override
    public Student login(String name, String pwd) {
       return studentMapper.login(name, pwd);
 }
```

步骤 5. 修改控制层 LoginController.java 源文件的 check 方法，根据不同的登录类型(type)，判断用户身份（学生、教师、管理员），以此分别调用不同的业务服务层方法。如果登录成功，则将 username 存储为 Session 对象，使之通过拦截器的过滤。代码如下：

```
    public  String  check(String  username,  String  password,  String  type,
HttpServletRequest request) throws DatabaseException {
        switch (Integer.parseInt(type)) {
            case 0: {
                Student student = studentServiceImpl.login(username, password);
                if (student != null) {
                    HttpSession session = request.getSession();
                    session.setAttribute("username", username);
                    return "redirect:/Student/list.do";
                } else {
                    return "login";
                }
            }
            case 1: {
                Teacher teacher = teacherServiceImpl.login(username, password);
                if (teacher != null) {
                    HttpSession session = request.getSession();
                    session.setAttribute("username", username);
                    return "redirect:/Teacher/list.do";
                } else {
                    return "login";
                }
            }
            case 2: {
                Admin admin = adminServiceImpl.login(username, password);
                if (admin != null) {
                    HttpSession session = request.getSession();
                    session.setAttribute("username", username);
                    return "redirect:/Admin/list.do";
                } else {
                    return "login";
                }
            }
            default:
            {
            return "login";
            }
        }
    }
```

步骤 6. 部署运行，访问 http://localhost:8077/login.jsp，分别用数据库表 student、teacher、admin 里的数据登录测试。如图 5-12 所示，使用教师姓名李青山测试登录功能，密码 0000，登录身份选择教师，单击“登录”按钮，则跳转到/Teacher/list.do 页面，如图 5-13 所示。

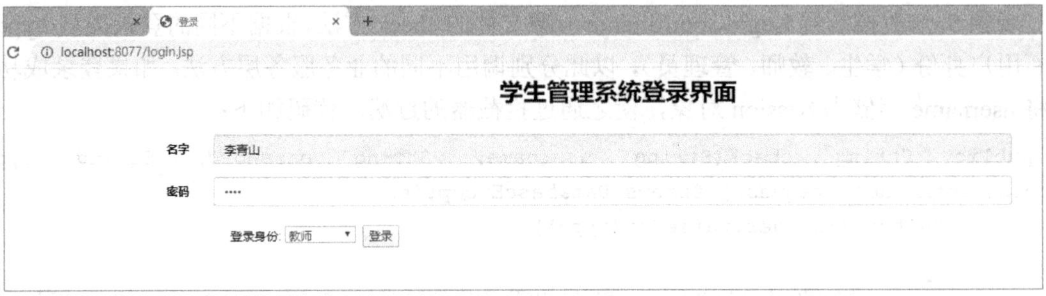

图 5-12　教师登录

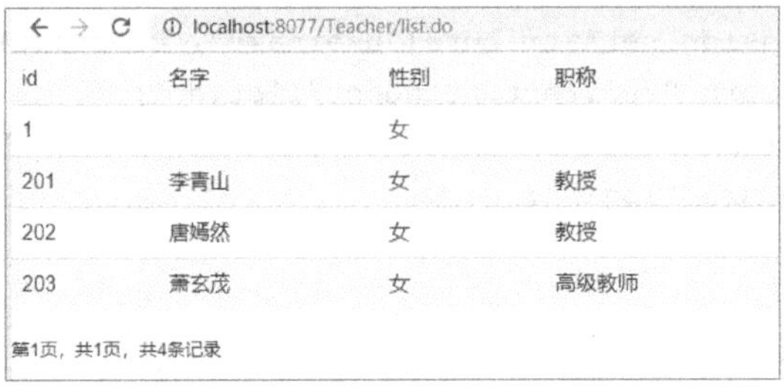

图 5-13　教师登录后 teacher 表数据显示效果

5.4.8　传入多个参数的写法

MyBatis 框架传入多个参数的写法有很多，还可以使用 Map、List 等集合类型。在 5.4.7 节的示例 5-5 中我们使用了下面的第二种写法。这里介绍三种常见写法。

1．使用索引的写法

```
public int  login(String name, String pwd);
<select id="login" resultType="BaseResultMap">
   select * from student where name= #{0} and pwd= #{1}
</select>
```

由于有多个参数，所以不能使用 parameterType（参数类型）来定位，改用#{index}表示索引的位置，从 0 开始。

2．使用注解的写法

```
Student login(@Param("name") String name,@Param("password") String password);

<select id="login" resultMap="BaseResultMap" parameterType="java.lang.String" >
     select
   <include refid="Base_Column_List" />
   from student
   where studentName = #{name} and password=#{password}
 </select>
```

这种写法最常见，方法头的入参跟 SQL 语句的#{参数名}保持一致，一目了然。

3．用 List 封装参数的写法

```
public List<Student> getStudentList(List<String> list);
<select id="getStudentList" resultType="XXBean">
```

```
    select 字段... from student where id in
     <foreach item="item" index="index" collection="list" open="(" separator=","
close=")">
         #{item}
    </foreach>
  </select>
```

getStudentList 方法的入参是 list，搜索 id 在 list 范围内的学生信息。假设 list 是['1','2','3','4']，foreach 最后的结果是 select 字段... from student where id in ('1','2','3','4')。

5.5 数据库连接技术

本节将会阐述 5 种数据库连接技术。其中，连接池技术在项目开发中使用最为频繁。连接池的基本思想是在系统初始化的时候，将数据库连接作为对象存储在内存中，当用户需要访问数据库时，并非建立一个新的连接，而是从内存的连接池中取出一个已建立的空闲连接对象。使用完毕后，也并非断开连接，而是将连接放回连接池中，以供下一个用户请求访问时使用。而连接的建立、断开都由连接池自身来管理。同时，还可以通过设置连接池的参数来控制连接池中的初始连接数、连接的上下限数及每个连接的最大使用次数、最大空闲时间等。也可以通过其自身的管理机制来监视数据库连接的数量、使用情况等。

Spring 在第三方依赖包中包含了两个数据库连接池的实现类包，其一是 Apache 的 DBCP，其二是 C3P0。可以在 Spring 配置文件中利用这两者中任何一个配置数据源。

5.5.1 DBCP

DBCP（DataBase Connection Pooling，数据库连接池）是 Apache 上的一个 Java 连接池项目，它是依赖 Jakarta commons-pool 对象池机制的数据库连接池，我们可以在 Maven 仓库里找到第三方包。

在 build.gradle 中添加如下依赖就可以使用。

```
// https://mvnrepository.com/artifact/org.apache.commons/commons-dbcp2
compile group: 'org.apache.commons', name: 'commons-dbcp2', version: '2.1.1'
```

下面是在 resources/spring-mybatis.xml 文件中使用 DBCP 配置 MySQL 数据源的片段：

```
  <!-- 数据库连接池 -->
    <bean id="dataSource" class="org.apache.commons.dbcp.BasicDataSource" destroy-
    method="close">
        <property name="driverClassName" value="${jdbc.driver}"/>
        <property name="url" value="${jdbc.url}"/>
        <property name="username" value="${jdbc.username}"/>
        <property name="password" value="${jdbc.password}"/>
  </bean>
```

BasicDataSource 提供了 close()方法来关闭数据源，所以必须设定 destroy-method="close" 属性，以便 Spring 容器关闭时，数据源能够正常关闭。除了以上必需的数据源属性，还有一些常用的属性，可以在 resources/jdbc.properties 文件中定义：

defaultAutoCommit：设置从数据源中返回的连接是否采用自动提交机制，默认值为 true。

defaultReadOnly：设置数据源是否仅能执行只读操作，默认值为 false。

maxActive：数据库最大连接数，设置为 0 时，表示没有限制。

maxIdle：最大等待连接数，设置为 0 时，表示没有限制。

maxWait：最大等待秒数，单位为毫秒， 超过时间会报出错误信息。

validationQuery：用于验证连接是否成功的查询 SQL 语句，SQL 语句必须至少返回一行数据，如可以简单地设置为“select count(*) from user”。

removeAbandoned：设置是否自我中断，默认值是 false。

removeAbandonedTimeout：设置经过几秒后数据连接会自动断开，在 removeAbandoned 为 true 时，提供该值。

logAbandoned：设置是否记录中断事件，默认值为 false。

5.5.2 C3P0 连接池

C3P0 是一个开放源代码的 JDBC 数据源实现项目，它实现了 JDBC 3 和 JDBC 2 扩展规范说明的 Connection 池和 Statement 池。在 Maven 仓库下搜索 C3P0 类包，使用下面的语句将其添加到依赖：

```
// https://mvnrepository.com/artifact/com.mchange/c3p0
compile group: 'com.mchange', name: 'c3p0', version: '0.9.5.2'
```

下面是在 spring-mybatis.xml 文件中使用 C3P0 配置数据源的片段：

```
<!-- 数据库连接池 -->
    <bean id="dataSource" class="com.mchange.v2.c3p0.ComboPooledDataSource" destroy-method="close">
        <property name="driverClass" value="${jdbc.driver}"/>
        <property name="jdbcUrl" value="${jdbc.url}"/>
        <property name="user" value="${jdbc.username}"/>
        <property name="password" value="${jdbc.password}"/>
</bean>
```

ComboPooledDataSource 提供了一个用于关闭数据源的 close()方法，这样就可以保证 Spring 容器关闭时数据源能够成功释放。C3P0 拥有比 DBCP 更丰富的配置属性，通过这些属性，可以对数据源进行各种有效的控制，同样可以在 resources/jdbc.properties 文件中定义以下属性：

acquireIncrement：设置当连接池中的连接用完时，C3P0 一次性创建新连接的数目。

acquireRetryAttempts：设置在从数据库获取新连接失败后重复尝试获取的次数，默认值为 30。

acquireRetryDelay：设置两次连接的间隔时间，单位为毫秒，默认值为 1000。

autoCommitOnClose：设置连接关闭时是否将所有未提交的操作回滚，默认值为 false。

automaticTestTable：C3P0 将建立一张名为 Test 的空表，并使用其自带的查询语句进行测试。如果定义了这个参数，那么属性 preferredTestQuery 将被忽略。程序员不能在这张 Test 表上进行任何操作，它只为 C3P0 测试所用，默认值为 null。

breakAfterAcquireFailure：获取连接失败将会引起所有等待获取连接的线程抛出异常，在这种情况下设置数据源是否仍有效保留，并在下次调用 getConnection()的时候继续尝试获取连接，默认值为 false。如果设为 true，那么在尝试获取连接失败后该数据源将申明已断开并永久关闭。

checkoutTimeout：设置当连接池用完时，客户端调用 getConnection()后等待获取新连接的时间，单位为毫秒，默认值为 0。超时后将抛出 SQLException，如设为 0 则无限期等待。

connectionTesterClassName：通过实现 ConnectionTester 或 QueryConnectionTester 的类来测试连接，类名需设置为全限定名，默认为 com.mchange.v2.C3P0.impl.DefaultConnectionTester。

idleConnectionTestPeriod：检查连接池中所有空闲连接的时长，默认值为 0，表示不检查。

initialPoolSize：设置初始化时创建的连接数，应在 minPoolSize 与 maxPoolSize 之间取值，默认值为 3。

maxIdleTime：设置最大空闲时间，超过空闲时间的连接将被丢弃。设置为 0 或负数则永不丢弃。默认值为 0。

maxPoolSize：设置连接池中保留的最大连接数。默认值为 15。

maxStatements：JDBC 的标准参数，用以控制数据源内加载的 PreparedStatement 数。但由于预缓存的 Statement 属于单个 Connection 而不是整个连接池，所以设置这个参数需要考虑多方面的因素。如果 maxStatements 与 maxStatementsPerConnection 均为 0，则缓存被关闭。默认值为 0。

maxStatementsPerConnection：连接池内单个连接所拥有的最大缓存 Statement 数。默认值为 0。

numHelperThreads：帮助线程数。由于 C3P0 是异步操作，缓慢的 JDBC 操作需要通过帮助进程完成。扩展这些操作可以有效地提升性能，通过多线程实现多个操作同时被执行。默认值为 3。

preferredTestQuery：定义所有连接测试都执行的测试语句。在使用连接测试的情况下设置这个参数能显著提高测试速度。测试的表必须在初始化数据源的时候就存在。默认值为 null。

propertyCycle：用户修改系统配置参数执行前最多等待的秒数。默认值为 300。

testConnectionOnCheckout：该参数如果设为 true，那么在每个 connection 提交的时候都将校验其有效性。这样，性能消耗大，故只在需要的时候才使用它。建议使用 idleConnectionTestPeriod 或 automaticTestTable 等方法来提升连接测试的性能。默认值为 false。

testConnectionOnCheckin：如果设为 true，那么在取得连接的同时将校验连接的有效性。默认值为 false。

5.5.3 获取 JNDI 数据源

JNDI（Java Naming and Directory Interface，Java 命名和目录接口）是 SUN 公司提供的一种标准的 Java 命名系统接口。当应用配置在高性能的应用服务器（如 WebLogic 或 Websphere 等）上时，我们可能更希望使用应用服务器本身提供的数据源。应用服务器的数据源使用开放的 JNDI 供调用者使用，Spring 为此专门提供引用 JNDI 资源的 JndiObjectFactoryBean 类。下面是一个简单的配置。

```
<bean id="dataSource" class="org.Springframework.jndi.JndiObjectFactoryBean">
    <property name="jndiName" value="java:comp/env/jdbc/bbt"/>
<bean>
```

上面的 jndiName 属性指定了引用的 JNDI 数据源名称"java:comp/env/jdbc/bbt"。

5.5.4 Spring 的数据源实现类

Spring 本身也提供了一个简单的数据源实现类 DriverManagerDataSource，它位于 org.Springframework.jdbc.datasource 包中。这个类实现了 javax.sql.DataSource 接口，但它并没有提供池化连接机制，每次调用 getConnection()获取新连接时，只是简单地创建一个新的连接。因此，这个数据源实现类比较适合在单元测试或简单的独立应用中使用，因为它不需要额外的依赖类。

5.5.5 Alibaba Druid

Alibaba Druid 据说是最快速的连接池技术，它结合了 C3P0、DBCP、Proxool 等数据库连接池的优点，增加了日志监控功能，同时它提供了强大的监控和扩展功能。它包括三个部分：DruidDriver（代理 Driver），能够提供基于 Filter-Chain 模式的插件体系；DruidDataSource（高效可管理的数据库连接池）；SQLParser（解析器）。使用前在 Maven 仓库中搜索 druid，添加如下依赖：

```
// https://mvnrepository.com/artifact/com.alibaba/druid
compile group: 'com.alibaba', name: 'druid', version: '1.1.10'
```

下面是 Alibaba Druid 数据库连接池在 resources/spring-mybatis.xml 文件中配置的片段，各标签和参数的意义请参看注释：

```
<!-- 阿里 druid 数据源 dataSource start -->
    <bean id="dataSource" class="com.alibaba.druid.pool.DruidDataSource"
          init-method="init" destroy-method="close">
        <property name="driverClassName" value="${jdbc.driver}" />
        <property name="url" value="${jdbc.url}" />
        <property name="username" value="${jdbc.username}" />
        <property name="password" value="${jdbc.password}" />
        <!-- 配置初始化大小、最小、最大 -->
        <property name="initialSize" value="10" />
        <property name="minIdle" value="10" />
        <property name="maxActive" value="50" />
        <!-- 配置等待获取连接的时间 -->
        <property name="maxWait" value="10000" />
        <!-- 配置间隔多久才进行一次检测，检测需要关闭的空闲连接，单位是毫秒 -->
        <property name="timeBetweenEvictionRunsMillis" value="60000" />
        <!-- 配置一个连接在池中最短生存的时间，单位是毫秒 -->
        <property name="minEvictableIdleTimeMillis" value="300000" />
        <property name="testWhileIdle" value="true" />
        <!-- 这里建议配置为 true，防止取到的连接不可用 -->
        <property name="testOnBorrow" value="true" />
        <property name="testOnReturn" value="false" />
        <!-- 打开 PSCache，并且指定每个连接上 PSCache 的大小 -->
        <property name="poolPreparedStatements" value="true" />
        <property name="maxPoolPreparedStatementPerConnectionSize"
                  value="20" />
        <!-- 这里配置提交方式，默认就是 true，可以不用配置 -->
```

```
    <property name="defaultAutoCommit" value="true" />
    <!-- 验证连接有效与否的 SQL，不同的数据配置不同  mysql:select 1  ;oracle : select 1
    from dual -->
    <property name="validationQuery" value="select 1 from dual" />
    <property name="filters" value="wall,stat" />
    <property name="proxyFilters">
        <list>
            <ref bean="logFilter" />
            <ref bean="stat-filter" />
        </list>
    </property>
</bean>
<!-- 慢 SQL 记录 -->
<bean id="stat-filter" class="com.alibaba.druid.filter.stat.StatFilter">
    <!-- 慢 SQL 时间设置，即执行时间大于 50 毫秒的都是慢 SQL -->
    <property name="slowSqlMillis" value="50"/>
    <property name="logSlowSql" value="true"/>
</bean>
<bean id="logFilter" class="com.alibaba.druid.filter.logging.Slf4jLogFilter">
    <property name="dataSourceLogEnabled" value="true" />
    <property name="statementExecutableSqlLogEnable" value="true" />
</bean>
<!--  阿里 druid 数据源 dataSource  end  -->
```

5.6 PageHelper 分页工具的使用

5.6.1 PageHelper 简介

PageHelper 是一款开源的 MyBatis 分页插件。它支持任何复杂的单表、多表分页；支持常见的 12 种数据库，如 Oracle、MySQL、MariaDB、SQLite、DB2 等；支持常见的 RowBounds（PageRowBounds）；使用 QueryInterceptor 规范。使用该插件实现分页显示更为轻松。

5.6.2 任务七：实现学生信息分页显示的功能

【示例 5-6】使用 PageHelper 实现学生信息分页显示。

步骤 1．在 build.gradle 中添加 pagehelper 依赖。代码如下：

```
//mvnrepository.com/artifact/com.github.pagehelper/pagehelper
compile group: 'com.github.pagehelper', name: 'pagehelper', version: '5.1.2'
```

步骤 2．修改 resources/spring-mybatis.xml 文件，在 sqlSessionFactory 的 plugins 的属性里添加 com.github.pagehelper.PageInterceptor 的 Bean 作为插件，并定义数据库语言 MySQL。代码如下：

```
<bean id="sqlSessionFactory" class="org.Mybatis.Spring.SqlSessionFactoryBean">
    <property name="dataSource" ref="dataSource" />
    <!-- 自动扫描 mapping.xml 文件 -->
```

```
    <property name="mapperLocations" value="classpath*:mapper/*.xml"></property>
  <!-- 配置分页插件 -->
  <property name="plugins">
  <array>
  <bean class="com.github.pagehelper.PageInterceptor">
     <property name="properties">
        <!--使用下面的方式配置参数，一行配置一个 -->
        <value>
           helperDialect=mysql
        </value>
     </property>
  </bean>
  </array>
  </property>
  </bean>
```

步骤 3．修改实现分页的类 StudentController.java 源码，实现 list 方法，在方法头的入参里使用@RequestParam(value="page",defaultValue= "1") int page 定义显示的页号，默认为第一页。代码如下：

```
  public String list(@RequestParam(value="page",defaultValue= "1") int page,  Model
model){
  //定义第几页和每页显示 5 条记录
  PageHelper.startPage(page,5);
     List<Student> students=studentServiceImpl.list();
  PageInfo pageInfo=new PageInfo(students);
  //通过 model 对象带出分页封装的 students 数据 pageInfo 到 listStudent.jsp 页面
  model.addAttribute("pageInfo",pageInfo);
     return "listStudents";
  }
```

步骤 4．修改显示分页数据的页面 listStudents.jsp，两个代码片段如下：

（1）在 table 的<c:forEach>标签中将 items 属性改为${pageInfo.list}，其他保持不变。

```
  <table class="table table-striped">
     <tr>
        <td>id</td>
        <td>名字</td>
        <td>性别</td>
        <td>班级</td>
        <td>生日</td>
        <td>头像</td>
        <td>编辑</td>
     </tr>
     <c:forEach items="${pageInfo.list}" var="student">
        <tr>
           <td>${student.id}</td>
           <td>${student.name}</td>
           <td>${student.sex==0?'男':'女'}</td>
           <td>${student.clazz}</td>
```

```
            <td>${student.birthday}</td>
            <td>${student.image}</td>
            <td><a href="delete.do?id=${student.id}">删除</a>
                <a href="toUpdate.do?id=${student.id}">修改</a>
            </td>
        </tr>
    </c:forEach>
</table>
```

（2）在 table 下面或上面加上分页的导航条。

```
<div class="row">
        <div class="col-md-6">
            第${pageInfo.pageNum}页，共${pageInfo.pages}页，共${pageInfo.total}条记录
        </div>
        <div class="col-md-6 offset-md-4">
            <nav aria-label="Page navigation example">
                <ul class="pagination pagination-sm">
                    <li class="page-item"><a class="page-link" href="list.do?page=1">
                    首页</a></li>
                    <c:if test="${pageInfo.hasPreviousPage}">
                        <li class="page-item"><a class="page-link"
                                                href="list.do?page=${pageInfo.pageNum-1}">
上一页</a></li>
                    </c:if>
                    <c:forEach items="${pageInfo.navigatepageNums}" var="page">
                        <c:if test="${page==pageInfo.pageNum}">
                            <li class="page-item active"><a class="page-link" href="#">
${page}</a></li>
                        </c:if>
                        <c:if test="${page!=pageInfo.pageNum}">
                            <li class="page-item"><a class="page-link"
                                                    href="list.do?page=${page}">${page}
</a></li>
                        </c:if>
                    </c:forEach>
                    <c:if test="${pageInfo.hasNextPage}">
                        <li class="page-item"><a class="page-link"href="list.do?page=
                                                ${pageInfo.pageNum+ 1}" >下一页</a></li>
                    </c:if>
                    <li class="page-item"><a class="page-link" href="list.do?page=
                    ${pageInfo.pages}">末页</a>
                    </li>
                </ul>
            </nav>
        </div>
    </div>
    </div>
```

步骤 5. 部署运行，访问 http://localhost:8077/Student/list.do，数据显示的 table 有分页功能，

运行效果如图 5-14 所示。

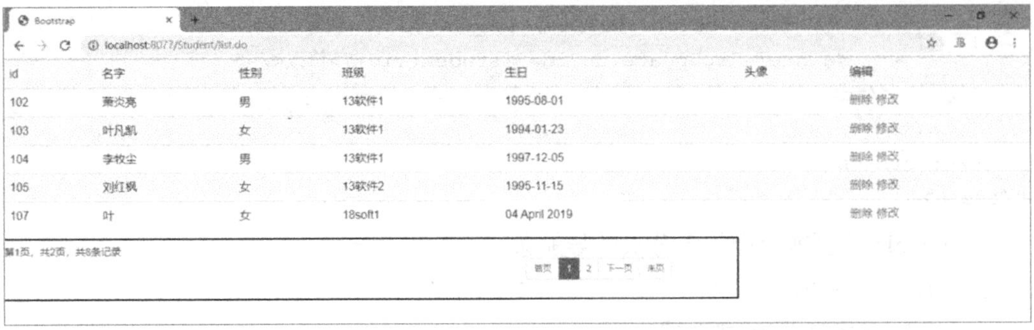

Bootstrap

localhost:8077/Student/list.do

id	名字	性别	班级	生日	头像	编辑
102	萧炎亮	男	13软件1	1995-08-01		删除 修改
103	叶凡凯	女	13软件1	1994-01-23		删除 修改
104	李牧尘	男	13软件1	1997-12-05		删除 修改
105	刘红枫	女	13软件2	1995-11-15		删除 修改
107	叶	女	18soft1	04 April 2019		删除 修改

第1页，共2页，共8条记录

首页 1 2 下一页 末页

图 5-14　分页效果

5.7　MyBatis 关联查询

有了前面几节的基础，我们可以处理一些简单的应用了。在实际项目中，关联表的查询很常见，如一对一、多对一、一对多及多对多。如何处理这种查询呢？这一节就专门讲这个问题。

5.7.1　任务八：实现一对一关系的处理

生活中，一对一关系广泛存在。以用户和订单举例，一个订单只属于一个用户，订单与用户是一对一关系；或者，一个人对应一张身份证，一张身份证对应一个人。

【示例 5-7】使用 association 标签实现一对一关系的处理。假设一个教师只能教授一门课程，则教师与课程之间就是一对一的关系。一对一的关系在映射文件中用 association 标签标注。

步骤 1．检查数据库中数据表：teacher、course。修改 resources/mapper/teacherMapper.xml 文件，首先改写 resultMap 标签，添加 association 子标签，使 teacher 与 course 建立一对一关联；其次增加 list 方法的 select 语句，使用等值连接使 teacher、course 两表产生关联。代码如下：

```
<resultMap id="BaseResultMap" type="model.Teacher">
  <!--
   WARNING - @mbggenerated
   This element is automatically generated by Mybatis Generator, do not modify.
  -->
  <id column="Id" jdbcType="INTEGER" property="id" />
  <result column="name" jdbcType="VARCHAR" property="name" />
  <result column="password" jdbcType="VARCHAR" property="password" />
  <result column="sex" jdbcType="INTEGER" property="sex" />
  <result column="birthday" jdbcType="VARCHAR" property="birthday" />
  <result column="course_id" jdbcType="INTEGER" property="courseId" />
  <result column="professional" jdbcType="VARCHAR" property="professional" />
  <association property="course" javaType="com.ssm.entity.Course">
    <id column="cid" property="id"/>
    <result column="cname" property="name"></result>
  </association>
```

```
</resultMap>
<select id="list"  resultMap="BaseResultMap">
  <!--
   WARNING - @mbggenerated
   This element is automatically generated by Mybatis Generator, do not modify.
  -->
  select
 t.*,c.id cid,c.name cname
  from teacher t ,course c
  where t.course_id=c.id
</select>
```

步骤 2．修改实体类 entity/Teacher.java 源码，添加 Course 对象属性，建立一对一关联。代码如下：

```
 private Course course;
 public Course getCourse() {
       return course;
    }
 public void setCourse(Course course) {
       this.course = course;
}
```

步骤 3．修改 DAO 接口 java/com.ssm.dao/TeacherMapper.java 源文件，增加 SQL 语句对应的方法，方法名 list 务必要与步骤 1 里的 select id 一致。代码如下：

```
 List<com.ssm.entity.Teacher> list();
```

步骤 4．修改业务接口 java/com.ssm.service/TeacherService.java 源文件，添加 list 方法的声明。代码如下：

```
public interface TeacherService {
   Teacher login(String name,String pwd);
   List<Teacher> list();
}
```

步骤 5．修改业务实现类 TeacherServiceImpl.java 源码，实现 list 方法，调用 DAO 接口的 list 方法。代码如下：

```
   @Override
   public List<Teacher> list() {
       return teacherMapper.list();
}
```

步骤 6．创建控制层的文件 java/com.ssm.controller/TeacherController.java，具体实现参照 StudentController.java 源文件。代码如下：

```
@Controller
@RequestMapping("/Teacher")
public class TeacherController {
   @Autowired
   private TeacherService  teacherServiceImpl;
   @RequestMapping("list.do")
   public String list(@RequestParam(value="page",defaultValue= "1") int page, Model
```

```
model){
        //定义第几页和每页显示的记录数
        PageHelper.startPage(page,5);
        //调用 studentServiceImpl 的 list 方法，从数据库取数据
        List<Teacher> teachers=teacherServiceImpl.list();
        //通过 model 对象带出分页封装的 students 数据 pageInfo 到 listStudent.jsp 页面
        PageInfo pageInfo=new PageInfo(teachers);
        model.addAttribute("pageInfo",pageInfo);
        return "listTeachers";
    }
}
```

步骤 7. 创建新文件 webapp/WEB-INF/views/listTeachers.jsp，用于展示 teacher 表数据。其中教授课程所对应的标签<td>${teacher.course.name}</td>体现了 teacher 与 course 之间的一对一关系。代码如下：

```
<%@page pageEncoding="UTF-8"%>
<%@taglib prefix="c" uri="http://java.sun.com/jsp/jstl/core" %>
<%@taglib prefix="Spring" uri="http://www.Springframework.org/tags" %>
<html>
<head>
    <meta charset="utf-8">
    <title> Bootstrap </title>
    <link href="/bootstrap/css/bootstrap.min.css" rel="stylesheet">
</head>
<body>
<!-- jQuery 文件。务必在 bootstrap.min.js 之前引入 -->
<script src="/bootstrap/js/jquery-3.3.1.min.js"></script>
<!-- 最新的 Bootstrap 核心 JavaScript 文件 -->
<script src="/bootstrap/js/bootstrap.min.js"></script>
<table class="table table-striped">
    <tr>
        <td>id</td>
        <td>名字</td>
        <td>性别</td>
        <td>职称</td>
        <td>生日</td>
        <td>薪水</td>
        <td>教授课程</td>
        <td>编辑</td>
    </tr>
    <c:forEach items="${pageInfo.list}" var="teacher">
        <tr>
            <td>${teacher.id}</td>
            <td>${teacher.name}</td>
            <td>${teacher.sex==0?'男':'女'}</td>
            <td>${teacher.professional}</td>
            <td>${teacher.birthday}</td>
            <td>${teacher.salary}</td>
```

```
                <td>${teacher.course.name}</td>
                <td><a href="delete.do?id=${teacher.id}">删除</a>
                    <a href="toUpdate.do?id=${teacher.id}">修改</a>
                </td>
            </tr>
        </c:forEach>
    </table>
    <div class="row">
        <div class="col-md-6">
            第${pageInfo.pageNum}页，共${pageInfo.pages}页，共${pageInfo.total}条记录
        </div>
        <div class="col-md-6 offset-md-4">
            <nav aria-label="Page navigation example">
                <ul class="pagination pagination-sm">
                    <li class="page-item"><a class="page-link" href="list.do?page=1">首
页</a></li>
                    <c:if test="${pageInfo.hasPreviousPage}">
                        <li class="page-item"><a class="page-link"
                                            href="list.do?page=${pageInfo.pageNum-1}">
上一页</a></li>
                    </c:if>
                    <c:forEach items="${pageInfo.navigatepageNums}" var="page">
                        <c:if test="${page==pageInfo.pageNum}">
                            <li   class="page-item   active"><a   class="page-link"   href=
"#">${page}</a></li>
                        </c:if>
                        <c:if test="${page!=pageInfo.pageNum}">
                            <li class="page-item"><a class="page-link"
                                                href="list.do?page=${page}">${page}</a>
</li>
                        </c:if>
                    </c:forEach>
                    <c:if test="${pageInfo.hasNextPage}">
                        <li class="page-item"><a class="page-link"
                                            href="list.do?page=${pageInfo.pageNum+1}">
下一页</a></li>
                    </c:if>
                    <li class="page-item"><a class="page-link" href="list.do?page= ${pageInfo.
pages}">末页</a></li>
                </ul>
            </nav>
        </div>
    </div>
    </div>
    </body>
    </html>
```

步骤 8．部署运行，访问 http://localhost:8077/Teacher/list.do，运行效果如图 5-15 所示。

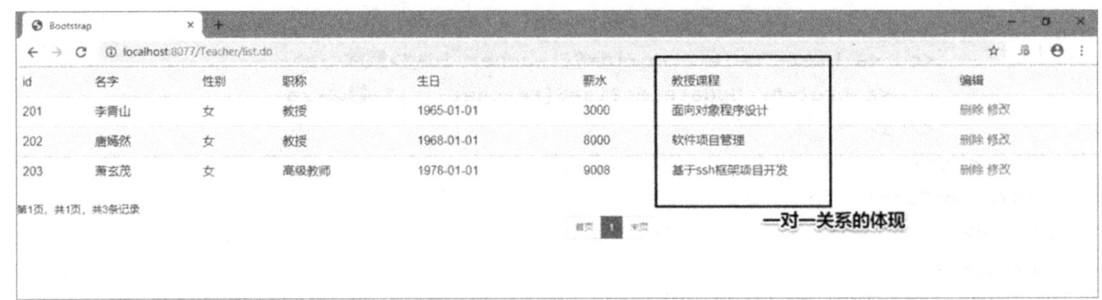

图 5-15　一对一关系显示效果

5.7.2　任务九：实现一对多关系的处理（三表联合查询）

一对多关系举例：一个用户可以拥有多个订单，用户和订单之间就是一对多关系。在附录 A 的数据库中，一个学生拥有多门课的成绩，student 表与 score 表之间就是一对多的关系。一对多关系可以使用 List 和 Set 集合类型来实现，两者在 MyBatis 框架的映射中都是通过 collection 标签加以实现的。

【示例 5-8】使用 collection 标签实现一对多关系的处理。同时，显示成绩所对应的课程名称。该示例涉及三表的联合查询。

步骤 1. 修改实体类 com.ssm.entity/Student.java，添加 List 类型的属性 scores 及 getter/setter 方法，体现 Student 类与 Score 类之间的一对多关系。代码如下：

```
private List scores;
public List getScores() {
    return scores;
}
public void setScores(List scores) {
    this.scores = scores;
}
```

步骤 2. 修改实体类 com.ssm.entity/Score.java，添加 Course 类型的属性 course，体现 Score 类与 Course 类之间的一对一关系。代码如下：

```
private Course course;
    public Course getCourse() {
        return course;
    }
    public void setCourse(Course course) {
        this.course = course;
    }
```

步骤 3. 修改 resources/mapper/ScoreMapper.xml 文件，添加对 Course 类的一对一关联。代码如下：

```
<association property="course" javaType="com.ssm.entity.Course">
    <result column="cname" property="name"></result>
</association>
```

步骤 4. 修改 resources/mapper/StudentMapper.xml 文件，使用 collection 添加对 Score 类的

一对多关联。代码如下：

```
<collection property="scores" ofType="com.ssm.entity.Score">
    <result column="course_id" jdbcType="INTEGER" property="courseId"/>
    <result column="score" jdbcType="DOUBLE" property="score"/>
    <association property="course" javaType="com.ssm.entity.Course">
      <result column="cname" property="name"></result>
    </association>
  </collection>
```

然后修改 list 方法的 SQL 语句。代码如下：

```
select st.*,c.name as cname from (select  student.*,score.course_id,score.score
from student
   LEFT OUTER JOIN score on student.id=score.student_id) st
   LEFT OUTER JOIN course c on st.course_id=c.Id
```

这里 student、score、course 三表的联合查询用到了 LEFT OUTER JOIN 左外连接。

步骤 5．修改 listStudents.jsp，显示列表里添加的学生选课名称和对应成绩。代码如下：

```
<c:forEach items="${student.scores}" var="sc">
      <c:out value="${sc.course.name}"/>
       <c:out value="${sc.score}"/><br/>
  </c:forEach>
```

步骤 6．部署运行，访问 http://localhost:8077/Student/list.do? page=1，效果如图 5-16 所示。

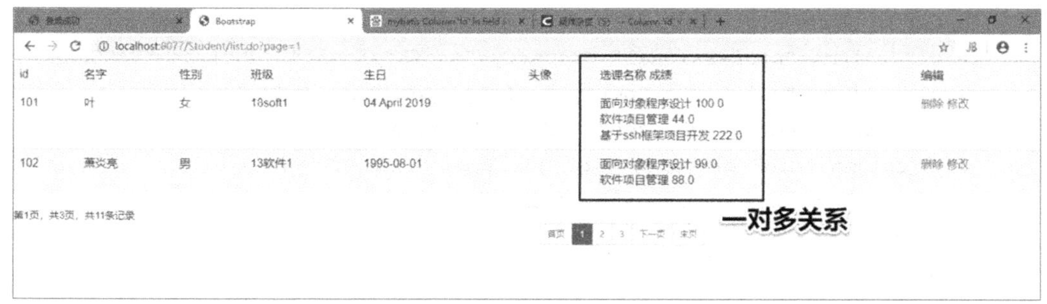

图 5-16　一对多关系显示效果

5.7.3　任务十：实现多对多关系的处理

生活中也有大量的多对多关系，如一个工程师可以属于多个项目团队，一个项目团队可以同时拥有多个工程师。

多对多关系的实现，即在双方的实体类里添加对方的 List 类型的属性，在映射文件 mapper.xml 中添加标签 collection 实现多对多关联。

我们继续使用数据库中的 teacher 和 course 表，实现一个教师教授多门课程，一门课程可以有多个教师授课的多对多关联情况。原本的数据库里，表 course 里有一个 teacher_id 字段，显示课程跟教师是一对一的关系，这里抛弃不用。重新创建一个新表 tbl_teacherCourse，实现它们之间的多对多关系。

【示例 5-9】

步骤 1．创建多对多关联表 tbl_teacherCourse。代码如下：

```
CREATE TABLE `test`.`tbl_teacherCourse` (
`id` int(11) NOT NULL AUTO_INCREMENT,
`teacherId` int(11) NULL DEFAULT NULL,
`courseId` int(11) NULL DEFAULT NULL,
PRIMARY KEY (`id`)
)
ENGINE=InnoDB
DEFAULT CHARACTER SET=utf8 COLLATE=utf8_general_ci
;
```

步骤 2．修改 Teacher 类、Course 类，为二者添加 List 类型属性，实现多对多的关联关系。代码如下：

```
Teacher ： private List<Course> courses;
Course: private List<Teacher> teachers;
```

并且生成属性 courses 和 teachers 的 getter/setter 方法。

步骤 3．修改 TeacherMapper.xml 文件，添加<collection>标签，说明教师与课程之间是一对多关系。代码如下：

```
<resultMap id="BaseResultMap" type="com.ssm.entity.Teacher" >
  <id column="Id" property="id" jdbcType="INTEGER" />
  <result column="name" property="name" jdbcType="VARCHAR" />
  <result column="password" property="password" jdbcType="VARCHAR" />
  <result column="sex" property="sex" jdbcType="INTEGER" />
  <result column="birthday" property="birthday" jdbcType="VARCHAR" />
  <result column="professional" property="professional" jdbcType="VARCHAR" />
  <result column="salary" property="salary" jdbcType="INTEGER" />
  <collection property="courses"  ofType="com.ssm.entity.Course">
    <id column="cid" property="id" jdbcType="INTEGER" />
  <result column="cname" property="name" jdbcType="VARCHAR" />
  </collection>
 </resultMap>
```

非常重要的是，还要在 TeacherMapper.xml 文件里添加一个 selectCourseById 的 SQL 语句的定义。代码如下：

```
<select id="selectCourseById" parameterType="int" resultMap="BaseResultMap">
  select t.*,c.id as cid,c.name as cname  from teacher t
  INNER JOIN tbl_teacherCourse tc on t.id=tc.teacherId
  INNER JOIN course c on c.id=tc.courseId where t.id=#{id,jdbcType=INTEGER}
 </select>
```

修改 CourseMapper.xml 文件，添加<collection>标签，说明课程与教师之间也是一对多关系。代码如下：

```
<resultMap id="BaseResultMap" type="com.ssm.entity.Course" >
  <id column="Id" property="id" jdbcType="INTEGER" />
  <result column="name" property="name" jdbcType="VARCHAR" />
  <result column="teacher_id" property="teacherId" jdbcType="INTEGER" />
  <collection property="teachers" ofType="com.ssm.entity.Teacher">
   <id column="tid" property="id" jdbcType="INTEGER" />
```

```
        <result column="tname" property="name" jdbcType="VARCHAR" />
      </collection>
    </resultMap>
```

非常重要的是，还要在 CourseMapper.xml 文件里添加一个 selectTeachersById 的 SQL 语句的定义。代码如下：

```
<select id="selectTeachersById" parameterType="int" resultMap="BaseResultMap">
    select c.*,t.id as tid,t.name as tname  from course c
    INNER JOIN tbl_teacherCourse tc on c.id=tc.courseId
    INNER JOIN teacher t on t.id=tc.teacherId where c.id=#{id,jdbcType=INTEGER}
  </select>
```

步骤 4. 修改 DAO 接口 com.ssm.dao/TeacherMapper.java 源文件，添加方法 selectCourseById，通过教师 id 查找该教师的信息及所授课程。代码如下：

```
Teacher selectCourseById(int id);
```

修改 DAO 接口 com.ssm.dao/CourseMapper.java 文件，添加方法 selectTeacherById，通过课程 id 查找该课程的信息及授课教师信息。代码如下：

```
Course selectTeachersById(int id);
```

步骤 5. 修改业务接口 com.ssm.service/TeacherService.java 文件，定义方法：

```
Teacher selectCoursesById(int id);
```

新建业务接口 com.ssm.service/CourseService.java 文件，定义方法：

```
public interface CourseService {
public Course selectTeachersById(int id);
public List<Course>  list();
}
```

步骤 6. 修改业务实现类 com.ssm.service/impl/TeacherServiceImpl.java 文件，实现 selectCoursesById 方法。代码如下：

```
public Teacher selectCoursesById(int id) {
        return teacherMapper.selectCourseById(id);
}
```

新建业务实现类 com.ssm.service/impl/CourseServiceImpl.java 文件，实现 selectTeachersById 方法。代码如下：

```
@Service
public class CourseServiceImpl implements CourseService {
   @Autowired
   private CourseMapper courseMapper;
   @Override
   public Course selectTeachersById(int id) {
       return courseMapper.selectTeachersById(id);
}
 @Override
   public List<Course> list() {
       return courseMapper.list();
   }
}
```

步骤 7．修改控制层 com.ssm.controller/TeacherController.java 文件，添加 selectCourseByid 方法。代码如下：

```
@RequestMapping("showCourses.do")
    public String selectCourseByid(int id,Model model){
        Teacher teacher=teacherServiceImpl.selectCoursesById(id);
        model.addAttribute("teacher",teacher);
        return "listTeacherCourse";
}
```

添加控制层 com.ssm.controller/CourseController.java 文件，其内容类似于 TeacherController.java。具体代码如下：

```
@Controller
@RequestMapping("/Course")
public class CourseController {
    @Autowired
    private CourseService courseServiceImpl;
    @RequestMapping("list.do")
    public String list(@RequestParam(value="page",defaultValue= "1") int page, Model
    model){
        //定义第几页和每页显示的记录数
        PageHelper.startPage(page,5);
        //调用 courseServiceImpl 的 list 方法，从数据库取数据
        List<Course> courses=courseServiceImpl.list();
        //通过 model 对象带出分页封装的 courses 数据 pageInfo 到 listCourse.jsp 页面
        PageInfo pageInfo=new PageInfo(courses);
        model.addAttribute("pageInfo",pageInfo);
        return "listCourses";
    }
    @RequestMapping("showTeachers.do")
    public String selectTeachersByid(int id,Model model){
        Course course=courseServiceImpl.selectTeachersById(id);
        model.addAttribute("course",course);
        return "listCourseTeacher";
    }
}
```

步骤 8．修改 webapp/WEB-INF/views/listTeachers.jsp 文件，在 table 标签的循环体里增加一个超链接 <a href="showCourses.do?id=${teacher.id}">所授课程</a>，单击超链接会跳转到显示该教师所授课程的页面。

创建新文件 webapp/WEB-INF/views/listCourses.jsp，内容与 listTeachers.jsp 文件类似，主体代码如下：

```
<table class="table table-striped">
    <tr>
        <td>id</td>
        <td>名字</td>
        <td>编辑</td>
    </tr>
    <c:forEach items="${pageInfo.list}" var="course">
        <tr>
```

```
            <td>${course.id}</td>
            <td>${course.name}</td>
            <td><a href="delete.do?id=${course.id}">删除</a>
                <a href="toUpdate.do?id=${course.id}">修改</a>
                <a href="showTeachers.do?id=${course.id}">授课教师</a>
            </td>
        </tr>
    </c:forEach>
</table>
```

步骤 9．创建新文件 webapp/WEB-INF/views/listTeacherCourse.jsp，用于显示每个教师所授的课程。主体代码如下：

```
<table class="table table-striped">
    <tr>
        <td>id</td>
        <td>course name</td>
    </tr>
    <c:forEach items="${teacher.courses}" var="course">
        <tr>
            <td>${course.id}</td>
            <td>${course.name}</td>
        </tr>
    </c:forEach>
</table>
```

创建新文件 webapp/WEB-INF/views/listCourseTeacher.jsp，用于显示每门课程的授课教师。主体代码如下：

```
<table class="table table-striped">
    <tr>
        <td>id</td>
        <td>teacher name</td>
    </tr>
    <c:forEach items="${course.teachers}" var="teacher">
        <tr>
            <td>${teacher.id}</td>
            <td>${teacher.name}</td>
        </tr>
    </c:forEach>
</table>
```

步骤 10．部署运行，在浏览器地址栏里输入 http://localhost:8077/Teacher/list.do，显示效果如图 5-17 所示。

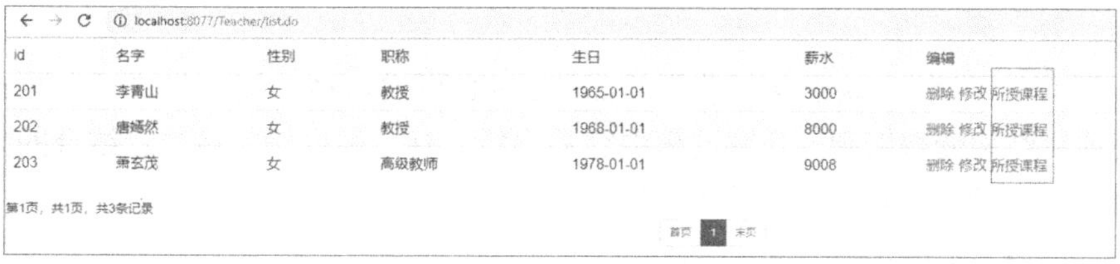

localhost:8077/Teacher/list.do

id	名字	性别	职称	生日	薪水	编辑
201	李青山	女	教授	1965-01-01	3000	删除 修改 所授课程
202	唐嫣然	女	教授	1968-01-01	8000	删除 修改 所授课程
203	萧玄茂	女	高级教师	1978-01-01	9008	删除 修改 所授课程

第1页，共1页，共3条记录

首页 1 末页

图 5-17　多对多关系显示效果

单击唐嫣然一行的“所授课程”超链接，则跳转到 http://localhost:8077/Teacher/showCourses.do?id=202，显示效果如图 5-18 所示。

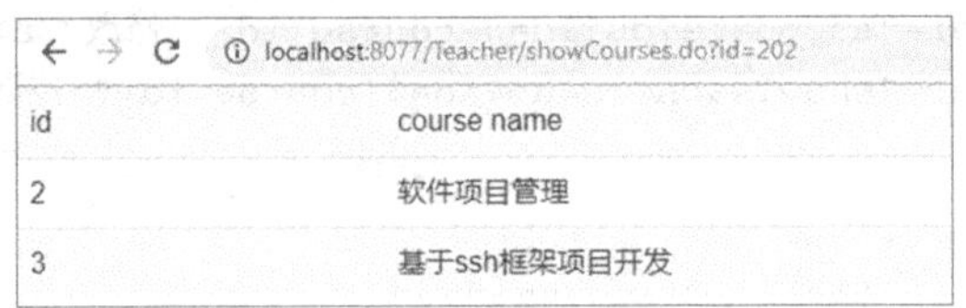

id	course name
2	软件项目管理
3	基于ssh框架项目开发

图 5-18　唐嫣然所授课程

在浏览器地址栏里输入 http://localhost:8077/Course/list.do，显示效果如图 5-19 所示。

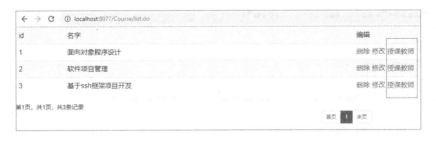

id	名字	编辑
1	面向对象程序设计	删除 修改 授课教师
2	软件项目管理	删除 修改 授课教师
3	基于ssh框架项目开发	删除 修改 授课教师

第1页，共1页，共3条记录

首页 1 末页

图 5-19　课程的授课教师

单击“软件项目管理”一行的“授课教师”超链接，则跳转到 http://localhost:8077/Course/showTeachers.do?id=2，显示教授软件项目管理课程的所有教师的 id 和姓名。显示效果如图 5-20 所示。

localhost:8077/Course/showTeachers.do?id=2

id	teacher name
201	李青山
202	唐嫣然

图 5-20　软件工程管理这门课的授课教师

经过上面 10 个步骤，我们完成了教师与课程之间多对多的关联映射，以及数据之间的简单呈现。我们重新梳理整个过程，首先是新建一个多对多的关联表，其次是在实体类添加 List 类型属性和在 mapper.xml 文件中使用 collection 标签建立多对多的关联关系。然后分别在 DAO 层、service 层、控制层实现对数据库的访问，以及最后在 JSP 文件中呈现出多对多的关系效果。

5.8　注解实现

在前面所有的示例中，映射文件使用的都是 XML 形式。DAO 接口与映射文件 mapper.xml 的分离，使项目结构变得繁杂，开发过程中容易出错。有一种更好的将二者结合的方式，那就是实现 SQL 语句注解 DAO，具体的形式如下：

```
public interface UserQueryMapper {
    @Select("SELECT * FROM USER WHERE id = #{id}")
```

```
    public User findUserById(int id) throws Exception;
}
```

上面是一个通过主键查询 user 表的例子，SQL 语句作为@Select 注解的参数出现，@Select 注解了 findUserById 方法。这样的形式，看上去更为简洁明了，开发过程也更简单。

那么，如何实现 SQL 语句注解 DAO 呢？我们要回到 MyBatis Generator 自动生成持久层，修改 resources/Mybatis/generatorConfig.xml 文件，将里面 type 属性的值（以前写的是 XMLMAPPER）改写成 ANNOTATEDMAPPER，表示生成注解形式的 DAO 层。

```
<javaClientGenerator
 targetPackage="${mapperPackage}" targetProject="${src_main_java}"type="ANNOTATE
DMAPPER"/>
```

改完后，重新运行任务 MybatisGenerate，生成新的 DAO 层和实体层。resources/mapper 目录及其下的映射文件不再生成。

如果要运行项目，需要重新在 DAO 层添加各个类的 login 或 list 方法。

例如，TeacherMapper.xml 文件中的 list 方法改为如下代码。

```
 @Select({
          "select",
          "t.*,c.name as cname",
          "from teacher t left outer join course c on t.course_id=c.id",
 })
 @Results({
          @Result(column="Id", property="id", jdbcType=JdbcType.INTEGER, id=true),
          @Result(column="name", property="name", jdbcType=JdbcType.VARCHAR),
          @Result(column="password",  property="password",  jdbcType=JdbcType.
VARCHAR),
          @Result(column="sex", property="sex", jdbcType=JdbcType.INTEGER),
          @Result(column="birthday",  property="birthday",  jdbcType=JdbcType.
VARCHAR),
          @Result(column="course_id",  property="courseId",  jdbcType=JdbcType.
INTEGER),
          @Result(column="professional", property="professional", jdbcType=JdbcType.
VARCHAR),
          @Result(column="salary",property="salary",jdbcType=JdbcType.INTEGER),
          @Result(column="cname",  property="course.name",  jdbcType=JdbcType.
VARCHAR)
  })
  List<Teacher> list();
```

另外，还要修改 StudentMapper.java 文件，在 deleteByPrimaryKey 方法前添加如下 SQL 注解。

```
@Delete({
     "delete from student",
     "where Id = #{id,jdbcType=INTEGER}"
  })
  int deleteByPrimaryKey(Integer id);
```

插入新的学生记录的 SQL 注解：

```
    @Insert({
        "insert into student (Id, name, ",
        "password, sex, clazz, ",
        "birthday, image)",
        "values (#{id,jdbcType=INTEGER}, #{name,jdbcType=VARCHAR}, ",
        "#{password,jdbcType=VARCHAR}, #{sex,jdbcType=INTEGER}, #{clazz,jdbcType=
VARCHAR}, ",
        "#{birthday,jdbcType=VARCHAR}, #{image,jdbcType=VARCHAR})"
    })
    int insert(Student record);
```

其他地方无须修改，运行效果也不会改变。修改后的源码参看配套教学资源第 5 章的“5.8 注解实现”。

拓展练习：在注解形式的 DAO 层下实现 StudentMapper.java 文件的 list 方法，使之支持多表关联的查询。

小 结

本章集中了大量的非常重要的实训，完成了持久层的自动生成，以及 Spring、MyBatis、Spring MVC 三大框架的集成。并且，实现了对数据库的增、删、改、查及多表关联访问。学完这一章，学生已经可以开始着手课程设计，独立完成 Java Web 项目的开发。在整个教学过程中，添加了很多实用性的小知识如数据库连接池技术、DAO 注解和分页查询，有助于学生知识面的扩展。

习 题

一、单选题

1. 下列（　　）不属于 MyBatis 架构。

   A. API 接口层　　B. 控制层　　C. 数据处理层　　D. 基础支撑层

2. MyBatis 是（　　）的升级版本。

   A. Ibatis　　B. Spring　　C. Hibernate　　D. Struts

3. MyBatis 的核心处理类是（　　）。

   A. Session　　B. SqlSession

   C. Servlet　　D. SessionStoreDirectory

二、填空题

1. MyBatis 项目下有一个工具项目，即________，方便开发者生成项目数据库的 Model、Mapper、DAO 持久层代码。

2. 连接池基本的思想是在系统初始化的时候，将________作为对象存储在内存中，当用户需要访问数据库时，并非建立一个新的连接，而是从连接池中取出一个已建立的________连接对象。

3．在 Java 中，一对多可以根据________和________集合类型来实现，两者在 MyBatis 中都是通过________标签加以实现的。

4．数据库使用________字符集不会出现中文乱码现象。

三、简答题

1．描述和 Hibernate 框架比较起来，MyBatis 框架更为适用的场景。

2．描述数据库连接池的作用和意义。

3．MyBatis 的 DAO 组件是如何生成并注入 Service 类中的？

综合实训

实训 1．根据资源 studentdb.sql 中 teacher 表结构，显示其中数据，如题图 5-1 所示。

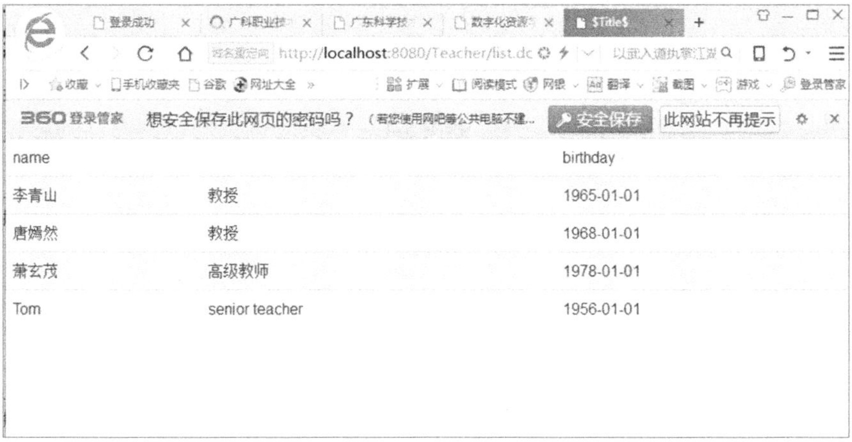
登录成功
$Title$
http://localhost:8080/Teacher/list.do
360登录管家 想安全保存此网页的密码吗？
安全保存
此网站不再提示
name
birthday
李青山
教授
1965-01-01
唐嫣然
教授
1968-01-01
萧玄茂
高级教师
1978-01-01
Tom
senior teacher
1956-01-01

题图 5-1

实训 2．根据登录视频，实现学生、教师、管理员三种身份的登录，如题图 5-2 所示。

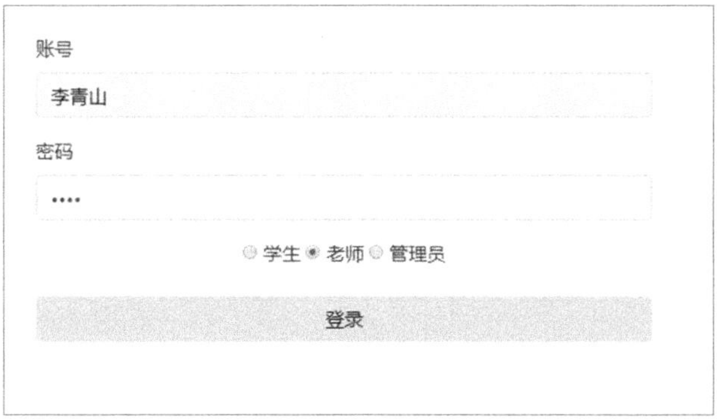
账号
李青山
密码
••••
学生
老师
管理员
登录

题图 5-2

实训 3．根据分页插件视频，完成 teacher 表数据的分页显示，如题图 5-3 所示。

实训 4．关联 student、course、score 表，在同一个页面显示每个学生每门课程的成绩，如题图 5-4 所示。

add teacher

id	name	password	sex	课程	birthday	operate
201	李青山	0000	1	面向对象程序设计	1965-01-01	update delete
202	唐嫣然	0000	1	软件项目管理	1968-01-01	update delete
203	萧玄茂	0000	1	基于ssh框架项目开发	1978-01-01	update delete
204	萧炎	0000	1	面向对象程序设计	1999-09-07	update delete

«上一页 1 2 3 下一页»

题图 5-3

add student

id	name	password	xingbie	clazz	birthday	scores	operate
101	林大雷	1111	female	13软件2	1996-03-24	面向对象程序设计 100.0 基于ssh框架项目开发 222.0	update delete
102	萧炎亮	0000	male	13软件1	1995-08-01	面向对象程序设计 99.0	update delete

第1页，共2页，共6条记录

首页 1 2 下一页 末页

题图 5-4

第 6 章

Spring IoC 在项目中的运用

本章学习目标

- 了解 Spring IoC 的原理
- 掌握 Spring IoC 基于 XML 和注解的管理和配置
- 掌握 Spring IoC 在项目中的运用

本章介绍了 Spring 框架的起源和 Spring 核心 IoC（Inverse of Control）的基础知识，并通过案例演示如何在 Spring 容器中创建和装配、管理和使用基于 XML 和注解两种方式的 Java Bean。

6.1 Spring 快速上手

6.1.1 Spring 概述

Spring 是分层的 Java EE 应用的轻量级开源框架，它是 Java 领域的第一个开源项目。Spring 框架最开始的部分是由 Rod Johnson 于 2000 年为伦敦金融界提供独立咨询业务时撰写的。

在经历了早期 J2EE 正统框架臃肿、低效，脱离现实的种种学院派做法后，一些程序开发人员组成了团队自愿拓展 Spring 框架，于 2003 年 2 月在 Sourceforge 上构建了一个项目。在 Spring 框架上工作了一年之后，这个团队在 2004 年 3 月发布了第一个版本（Spring 1.0）。这个版本凭借优秀的文档功能和详尽的参考文献，还有不俗的理念“好的设计优于具体实现，代码必须易于测试”，在 Java 社区里变得异常流行。Spring 框架的一个重要设计目标就是更容易与已有的 J2EE（现在称之为 Java EE 或 JEE）标准和商用工具整合。简单地说，Spring 最初的目标，并不是致力于打造一个大而全的新框架，而是希望 Spring 框架成为一个项目黏合剂，能够快速方便地集成各种应用中的各种现有技术。开发人员在实现过程中使用各种框架感觉就像使用简单的 JavaBean 一样，在必要的时候还能轻松完成同类框架和工具的替换。而 Spring 达到这一目标的两大核心就是控制反转/依赖注入（Inversion of Control / Dependency Injection，IoC/DI）和面向切面编程（Aspect Oriented Programming，AOP）。其中，IoC/DI 是 Spring 最基本的底层；而 AOP 是 Spring 最大的亮点，它们也是我们学习 Spring 的起点。

在诞生之初，Spring 就是为了简化 EJB 中间件，替代日益重量级的企业级 Java EE 技术而出现的。但随着时间的推移，Java EE 自身也在不断地改进，面向简单 JavaBean 模型、依赖注入、面向切面思想和技术也被逐渐包容，这无疑是受到了 Spring 成功的启发。

在被模仿的同时，Spring 以其扎实的理论继续在其他领域不断拓展，移动开发、社交 API 集成、安全管理、NoSQL 数据库、云计算和大数据等都是它正在涉足的领域，其应用前景非常广阔。我们可以看到，Spring 已经打造出一个自己专属的 Spring 生态帝国（Spring 全家桶）。在 Spring 这里，拥有几乎所有 Web 开发所需要的解决方案。

Spring 框架以反转控制和面向切面编程为内核，提供了展现层 Spring MVC、持久层 Spring JDBC 及业务层事务管理等众多的企业级应用技术。并且，它整合了众多的第三方框架和类库，逐渐成为使用最多的 Java EE 企业级应用开源框架。

Spring 一直致力于最简洁的开发实现与测试，它给我们带来了许多意想不到的便利。

- 方便解耦，易于开发：通过 Spring IoC 容器，将对象间的依赖关系交于 Spring 控制，避免硬编码造成对程序的过度依赖。自从有了 Spring，开发人员就摆脱了传统的单例模式所带来的重复劳动。
- 支持 AOP 编程：通过 Spring AOP 功能，用户可以灵活方便地实现面向切面编程，它替代传统的面向对象编程。
- 支持声明式事务：用户不必使用重复单调的事务管理代码，通过 AOP 的声明式事务可以灵活地进行事务管理，提高开发效率和质量。
- 测试方便。
- 便于集成其他框架。
- 降低了 Java EE API 的使用难度。

Spring 框架是一个分层架构，由 1400 多个类构成，由 7 个定义良好的模块组成，如图 6-1 所示。Spring 模块构建在核心容器之上，核心容器定义了创建、配置和管理 Bean 的方式。

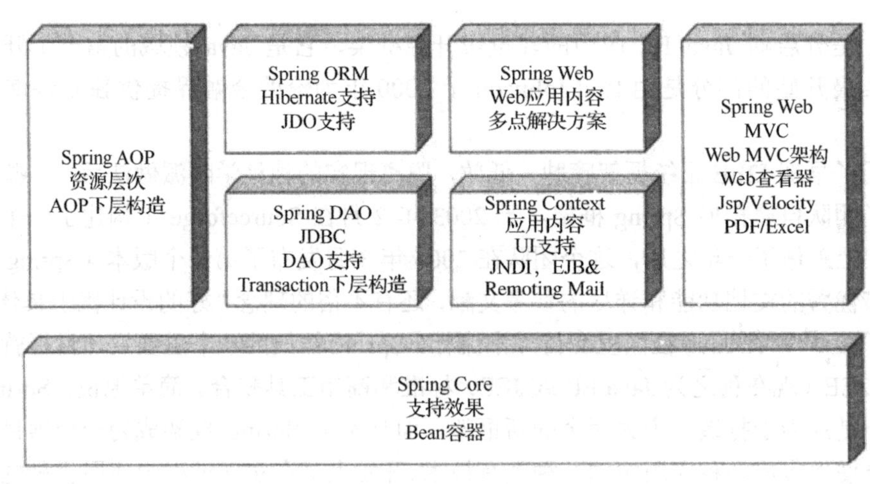

图 6-1　Spring 模块

Spring Core 模块是 Spring 框架的核心和基础，包括 IoC 和依赖注入功能。其中的 Bean 容器实现了 BeanFactory，它可以把配置和依赖从实际编码逻辑中分离。

Spring Context 模块在 Spring Core 模块的基础上建立起来，它以一种类似于 JNDI 注册的方式来访问对象。它添加了国际化事件传播、资源加载和透明地创建上下文（如通过 Servlet 容器）等功能。Spring Context 模块也支持 Java EE 的功能，如 EJB、JMX 和远程调用等。ApplicationContext 接口是 Spring Context 模块的核心。

Spring 数据访问层包括 Spring DAO 模块和 Spring ORM 模块。Spring DAO 模块包括 JDBC、DAO 支持和 Transaction 下层构造。JDBC 提供了 JDBC 抽象层，它大大减轻了 JDBC 编码工作量。Spring ORM 模块是对象关系映射 API 的集成，包括 JPA、JDO 和 Hibernate 等。此模块可以整合 ORM 框架和 Spring 的其他功能，如事务管理，可实现特殊接口类及所有的 POJO 支持的编程式和声明式事务管理。

Spring Web 层由 Spring Web 模块和 Spring Web MVC 模块组成。Spring Web 模块提供面向 Web 的基本功能和面向 Web 的应用上下文，如多媒体的文件上传功能，使用 Servlet 监听器初始化 IoC 容器等。它还包括 HTTP 客户端以及 Spring 远程调用中与 Web 相关的部分。Spring Web MVC 模块为 Web 应用提供了模型视图控制（MVC）和 REST Web 服务的实现。其中，Spring 的 MVC 框架可以使业务代码和用户交互表单分离，且可以集成 Spring 框架的其他所有功能。

Spring AOP 模块提供了面向切面的编程实现，实现了使用拦截器和切点对代码进行干净的解耦，使实现功能的代码与业务代码彻底解耦。Spring AOP 是一个功能强大且成熟的面向切面编程框架。

6.1.2 Spring IoC 依赖

项目添加 Spring IoC 的依赖，只需要引入 spring-core.jar 及 spring-beans.jar 文件就可以了。它们包含了访问配置文件、创建和管理 JavaBean 以及进行 IoC/DI 操作相关的所有类。从第 4 章开始，项目构建文件 build.gradle 中添加的依赖 spring-web 包里就包含了 spring-core.jar 及 spring-beans.jar 文件。所以，项目并不需要额外添加依赖。

6.2 Spring 的核心技术——控制反转 IoC

6.2.1 IoC 思想概述

传统设计中，假设一架飞机的设计流程可能是这样的：先设计机翼，然后根据机翼大小设计机身，接着根据机身设计发动机，最后根据发动机设计整个飞机。这里就出现了一个“依赖”关系：飞机依赖发动机，发动机依赖机身，机身依赖机翼。这样的设计看起来没问题，但是成品的可维护性非常差。当设计完工之后，如果中美之间发生了贸易摩擦，飞机市场需求发生了巨大的变化，需要我们更改飞机的机翼尺寸。这下我们的麻烦就来了，因为根据机翼设计的机身、发动机都必须跟着改。

控制反转（Inverse of Control，IoC）的思想就是换一种思路，颠倒设计的过程。我们先设计飞机的整体模样，然后根据飞机设计发动机，再根据发动机设计机身，最后根据机身设计机翼。这样，整体与局部的依赖关系就倒置过来：机翼依赖机身， 机身依赖发动机，发动机依赖飞机。

IoC 是由容器控制业务对象之间的依赖关系的，而非传统实现中，由程序代码直接操控依赖关系。这也正是所谓“控制反转”的概念所在：控制权由应用代码转到了外部容器，控制权的转移即反转。控制权转移带来的好处就是降低了业务对象之间的依赖程度。全部对象的控制权全部上缴给IoC容器，所以IoC容器成了整个系统的核心。它起到了一种类似黏合剂的作用，把系统中的所有对象黏合在一起发挥作用。如果没有这个“黏合剂”，对象与对象之间会彼此失去联系，这就是有人把 IoC 容器比喻成“黏合剂”的由来。

IoC 思想源于 DI（Dependency Injection，依赖注入），即组件之间的依赖关系由容器在运行期决定，即由容器动态地将某个依赖关系注入组件之中。依赖注入的目的并非为软件系统带来更多功能，而是为了提升组件重用的频率，并为系统搭建一个灵活、可扩展的平台。通过依赖注入机制，我们只需要通过简单的配置，而不需要任何代码就可指定目标需要的资源，完成自身的业务逻辑，而不需要关心具体的资源来自何处，由谁实现。

理解 DI 的关键是“谁依赖谁，为什么需要依赖，谁注入谁，注入了什么”。

谁依赖于谁：应用程序依赖于 IoC 容器。

为什么需要依赖：应用程序需要 IoC 容器来提供对象需要的外部资源。

谁注入谁：IoC 容器注入应用程序依赖的对象。

注入了什么：注入某个对象所需要的外部资源（包括对象、资源、常量数据）。

相对 IoC 而言，依赖注入更加准确地描述了这种古老而又时兴的设计理念。

用一句话概括依赖注入就是“不要来找我，我会去找你”，一个类不需要去查找或实例化它们所依赖的类。对象间的依赖关系是在创建对象时由负责协调项目中各个对象的外部容器来提供并管理的。也就是强调了对象间的某种依赖关系是由容器在运行期间注入调用者的，控制程序间关系的实现交给了外部的容器来完成。这样，当调用者需要被调用对象时，调用者不需要知道具体实现细节，而只需要从容器中拿出一个对象并使用就可以了。

6.2.2 Spring IoC 实现

1. 容器

Spring IoC 是一个实现了控制反转思想的容器，它通过配置文件 beans.xml 描述 Bean 之间的依赖关系，利用 Java 语言的反射（Reflection）机制实例化 Bean 并建立 Bean 之间的依赖。Spring IoC 容器除了实现上面所述的底层工作，还提供了 Bean 实例缓存、生命周期管理、Bean 实例代理等高级事务。

BeanFactory 是 Spring 框架最核心的接口，它提供了高级 IoC 的配置机制。BeanFactory 是一个非常纯粹的 Bean 容器，它可以创建并管理各种类的对象。

而 ApplicationContext 称为应用上下文，是 BeanFactory 的子接口，有时也被称为 Spring 容器。Spring 的应用者面对的是 ApplicationContext 接口的通用工厂，可以创建并管理各种类的对象。这个大名鼎鼎的 Spring 容器，与我们的应用项目息息相关，它继承了 BeanFactory，所以它是 BeanFactory 的扩展升级版。ApplicationContext 的结构决定了它与 BeanFactory 的不同，其主要区别有：

（1）继承 MessageSource，提供国际化的标准访问策略。

（2）继承 ApplicationEventPublisher，提供强大的事件机制。

（3）扩展 ResourceLoader，可以用来加载多个 Resource，可以灵活访问不同的资源。

（4）对 Web 应用的支持。

2．容器管理的 Bean

所谓 Bean，就是一个 POJO（普通 Java 对象），Spring 称这些被创建和被管理的对象为 Bean。容器需要知道如何创建 Bean，如何管理 Bean 的生命周期，以及 Bean 之间的依赖关系。

过去用户使用 Spring 之前，必须在 Spring IoC 容器中装配好 Bean，并建立好 Bean 和 Bean 之间的关联关系。现在已经通过逐步的改进使得配置日益简洁优雅。

IoC 容器现在既可以使用 XML 配置，也可以使用注解配置，新版本的 Spring 还可以零配置实现 IoC。Spring 容器在初始化时先读取配置文件，根据配置文件或元数据创建与组织对象并存入容器中，当程序需要使用对象时再从 IoC 容器中取出需要的对象。具体的 IoC 容器启动和创建 Bean 可以参看图 6-2。

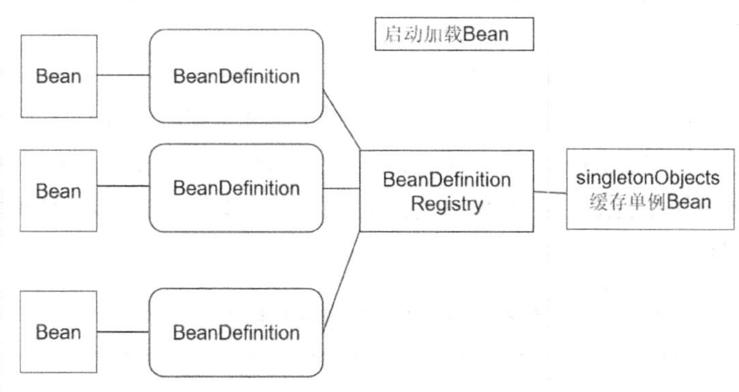

图 6-2 IoC 容器启动和创建 Bean

Spring IoC 容器启动初期，会为所有配置过的 Bean 对象都各自生成一个 BeanDefinition 的定义表，存储到 BeanDefinitionRegistry 的 BeanDefinitionMap 中。用户通过 BeanFactory 接口的实现类如 ClassPathXmlApplicationContext 去获取 Bean 对象的时候，会从 BeanDefinitionMap 中获取这个 Bean 的定义 BeanDefinition，然后通过反射生成一个 Bean。默认为单例时，还会把这个实例化的 Bean 存储在单例对象列表 singletonObjects 中，下次再获取 Bean 对象的时候就直接从 singletonObjects 中返回。

采用 XML 方式配置 Bean 的时候，Bean 的定义信息和实现是分离的；而采用注解的方式可以把两者合为一体，Bean 直接以注解的形式定义在实现类中，从而达到零配置的目的。

Bean 的配置信息是 Bean 的元数据信息，它由以下 4 个部分组成：

- Bean 的实现类。
- Bean 的属性信息。
- Bean 的依赖关系。
- Bean 的行为配置。

Bean 的配置信息文件定义了 Bean 的声明和依赖关系。Spring IoC 容器会根据 Bean 的配置信息在容器内部建立 Bean 的定义注册表，然后根据注册表加载、实例化 Bean，并建立 Bean 之间的依赖关系，最后将这些准备就绪的 Bean 放到 Bean 缓存池中，以供外层的应用程序调用。Spring 管理这些 Bean 的生命周期的主要过程：实例化 Bean 执行顺序，调用构造函数和属性值的 set 方法、init 初始化方法（如果配置了）、destory（销毁）方法。

下面是基于 XML 配置的信息文件格式。

```
<?xml version="1.0" encoding="UTF-8"?>
< beans    xmlns=http://www.Springframework.org/schema/beans
xmlns:xsi=http://www.w3.org/2001/XMLSchema-instance
xsi:schemaLocation="http://www.Springframework.org/schema/beans
http://www.Springframework.org/schema/beans/Spring-beans-2.5.xsd">
< bean id="……"     class="……">
  ……
</bean>
</beans>
```

<beans>是 XML 配置文件中最顶层的元素，它下面可以包含 0 或者 1 个<description>和多个<bean>及<import>或者<alias>。

其中，id 是这个 Bean 的名称，id 在 IoC 容器中必须唯一，还必须满足 XML 对 id 的命名规范。如果没有指定 id，Spring 容器会自动将全限定类名作为 Bean 的名称。开发者可以通过容器实例化上下文对象的 getBean()方法获取对应的 Bean。从容器中取回的对象默认是单例模式的。class 属性指定了 Bean 对应的实现类。省略号里可以定义 Bean 的属性信息或者依赖关系等。更加详细的属性描述参见表 6-1。

表 6-1　<bean>元素的常用属性

属性名称	描　述
id	Bean 的唯一标识符
class	指定 Bean 的具体实现类，必须使用类的全限定名
name	与 id 类似，可以指定多个名称，每个名称之间用逗号或分号隔开
scope	Bean 实例的作用域，其属性值有 singleton、prototype、request、session、global session。默认值是 singleton
constructor-arg	使用构造参数实例化 Bean，该元素的 index 指定构造参数序号，type 指定构造参数的类型
property	调用 Bean 实例的 set 方法赋值，实现依赖注入。name 属性指定 Bean 对应的属性名
ref	该元素的 bean 属性指定引用的 Bean 名称
value	指定常量值
list	封装 List 或者数组类型的注入
set	封装 Set 类型属性的注入
map	封装 Map 类型属性的注入
entry	设置键值对。key 指定字符串类型的键值，ref 或 value 指定值

Spring IoC 容器对 Bean 的创建和管理方法有很多种，下面介绍基于 XML（6.3 节）和基于注解（6.4 节）及零配置（6.5 节）3 种方式。

6.3　基于 XML 的实例化 Bean

下面通过示例来详细解释基于 XML 的 Bean 的声明、配置和访问。

6.3.1　任务一：实现属性注入的 Bean 实例化

属性注入即通过 setXxx()方法将属性值或依赖对象注入 Bean。由于属性注入方法灵活，因此属性注入是实际应用中最常采用的注入方式。

【示例 6-1】声明一个小汽车 Car 的 Bean，通过<property>标签注入属性 type（车型）和 maxSpeed（最高速度）的值。

步骤 1. 创建 Car 类：属性 brand、maxSpeed，以及其 getter/setter 方法。另外增加一个 introduce 方法，用于在控制台输出 Bean 的属性值。代码如下：

```
public class Car {
    private String type;
    private int maxSpeed;
    public String getType() {
        return type;
    }
    public void setType(String type) {
        this.type = type;
    }
    public int getMaxSpeed() {
        return maxSpeed;
    }
    public void setMaxSpeed(int maxSpeed) {
        this.maxSpeed = maxSpeed;
    }
    public void introduce(){
        System.out.println("type is: "+ getType()+" ,maxSpeed is: "+getMaxSpeed());
    }
}
```

步骤 2. 撰写 Spring IoC 容器的配置文件 beans.xml，定义一个 Car 类型的 Bean，id 是 tom，为属性 type（车型）赋值 nissan，为属性 maxSpeed（最高速度）赋值 300。代码如下：

```
<?xml version="1.0" encoding="UTF-8"?>
<beans xsi:schemaLocation="http://www.Springframework.org/schema/beans http://www.
Springframework.org/schema/beans/Spring-beans-3.1.xsd"
              xmlns:p="http://www.Springframework.org/schema/p"
              xmlns:xsi="http://www.w3.org/2001/XMLSchema-instance" xmlns="http://
              www.Springframework.org/schema/beans">
    <bean id="tom" class="com.ssm.util.Car">
        <property name="type" value="nissan"/>
        <property name="maxSpeed" value="300"/>
    </bean>
</beans>
```

步骤 3. 在 Car 类中添加 JUnit 的单元测试方法，从类路径获取 beans.xml 后，调用 getBean 方法获取 tom bean，然后调用 car 的 introduce 方法在控制台输出 tom 对象的属性信息。代码如下：

```
@Test
    public void test(){
```

```
        ApplicationContext  applicationContext=new  ClassPathXmlApplicationContext
("beans.xml");
                Car car= (Car) applicationContext.getBean("tom");
        car.introduce();
    }
```

上面程序里，有一个重要的类 ApplicationContext，我们可以把它当作 Spring IoC 容器，这个容器通过实例化 ClassPathXmlApplicationContext 类读取位于类路径下的 beans.xml 文件，创建并管理里面的 Bean。

beans.xml 文件也可以放在普通的文件系统路径下，Spring IoC 容器可以通过实例化 FileSystemXmlApplicationContext 读取。

在获取 ApplicationContext 实例化对象 applicationContext 后，就可以像 BeanFactory 一样调用 getBean(beanName)返回 Bean 了。因为 getBean 方法返回 Object 类型的对象，所以在前面加了一个“(Car)”实现类型的强制转换。

ApplicationContext 会在初始化上下文时就实例化所有单例的实例，所以启动时间会有点长。

步骤 4．运行 test 方法，控制台上会输出小汽车 Bean“tom”的信息：

```
type is:nissan,maxSpeed is:200
```

6.3.2 任务二：实现构造方法注入的 Bean 实例化

如果上述 Car 类的所有对象在实例化时都必须提供 type 和 maxSpeed 值，使用属性注入方式就不能实现这个要求。这时通过构造方法注入可以满足这一要求。使用构造方法注入的前提是 Bean 必须提供带参的构造方法。

这里有一点要注意，Java 语言规定如果类中没有构造方法，会自动生成一个默认的无参构造方法。反之，如果已定义了构造方法，则 JVM 不会为其生成默认的构造方法，所以需要同时提供一个无参构造方法，否则使用属性注入时将会抛出异常。

让我们回到构造方法注入的具体实现。在有些时候，容器加载 XML 配置的时候，因为某些原因，会无法明确配置项与对象的构造方法参数列表的一一对应关系，这就需要用<constructor-arg>的 type（类型）或者 index（索引）属性加以标明。

【示例 6-2】按类型匹配入参。

为 Car 类增加一个可设置 type 和 maxSpeed 属性的构造方法。代码如下：

```
package com;
public class Car{
......
 public Car(String type, int maxSpeed) {
        this.type = type;
        this.maxSpeed = maxSpeed;
    }......
}
```

构造方法注入的配置方式和属性注入方式有所不同。按照 Spring IoC 容器的配置格式，要通过构造方法注入方式为当前业务对象注入其所依赖的对象，需要使用<constructor-arg>标签。

```
<bean id="tom" class="com.ssm.util.Car">
      <constructor-arg type="java.lang.String" value="nissan"/>
      <constructor-arg type="int" value="200"/>
</bean>
```

也可以使用索引的方式来注入，1 表示第一个属性 type，2 表示第 2 个属性 maxSpeed。代码如下：

```
<bean id="tom" class="com.ssm.util.Car">
      <constructor-arg index="1" value="nissan"/>
      <constructor-arg type="2" value="200"/>
</bean>
```

但是为了避免某些可能的歧义和错误，在实际的编码中，我们还是强烈建议使用 name 指定属性名来注入属性值，代码如下：

```
<bean id="tom" class="com.ssm.util.Car">
      <constructor-arg name="type" value="nissan"/>
      <constructor-arg name="maxSpeed" value="200"/>
</bean>
```

这才是一个更为良好的配置习惯。具体代码请大家参看配套教学资源中 6-2 源码。

6.3.3　任务三：实现 Bean 的引用

Spring 配置文件中，用户可以注入基本数据类型，也可以注入集合、Map 等类型，还可以让 Bean 之间互相引用。

1．基本数据类型

基本数据类型可以通过<value>标签进行注入。在前面的所有例子中，使用的都是基本数据类型属性注入。

2．对象类型

Spring IoC 容器中定义的 Bean 可以相互引用。

【示例 6-3】演示 Human 和 Car 两个 Bean 之间如何互相引用。

步骤 1. 构造一个新的类 Human，其中有一个属性 car 是 Car 类型的对象，还有一个是 String 类型的属性 name。除了两个属性的 getter/setter 方法，还有一个 introduce 方法，除用于输出 name 值，还调用了 car 的 introduce 方法输出 car 的信息。代码如下：

```
public class Human {
    private String name;
    private Car car;
    public String getName() {
        return name;
    }
    public void setName(String name) {
        this.name = name;
    }
    public Car getCar() {
        return car;
    }
    public void setCar(Car car) {
```

```
        this.car = car;
    }
    public void introduce(){
        System.out.println(getName()+" , car:");
        car.introduce();
    }
}
```

步骤 2．修改 beans.xml 文件，根据 Human 类创建一个新的 Bean，id 是 zhangsan，使用<ref>元素来给属性 car 注入值 tom。tom 是前面创建的 Car 类的对象。代码如下：

```
<bean id="zhangsan" class="com.ssm.util.Human">
    <property name="name" value="zhangsan"/>
    <property name="car" ref="tom"/>
</bean>
```

步骤 3．修改 test 方法，获取 id 为 zhangsan 的 Bean，并调用其 introduce 方法输出属性值。代码如下：

```
public void test(){
    ApplicationContext   applicationContext=new   ClassPathXmlApplicationContext
("beans.xml");
    Human zhangsan= (Human) applicationContext.getBean("zhangsan");
    zhangsan.introduce();
}
```

步骤 4．运行 test 方法，在控制台上输出属性值。代码如下：

```
zhangsan , car:
type is: nissan ,maxSpeed is: 200
```

<ref>元素除了示例 6-3 中的写法，还可以使用下面的写法：

```
<bean id="zhangsan" class="com.ssm.util.Human">
        <property name="name" value="zhangsan"/>
        <property name="car" >
        <ref bean="tom"/>
    </property>
    </bean>
```

<ref>元素可以使用下面 3 个属性来引用容器中的其他 Bean。

bean：该属性可以引用同一容器或父容器中的 Bean，这是最常用的方式。

local：该属性只能引用同一配置文件中定义的 Bean。

parent：该属性引用父容器中的 Bean。

3．集合类型

集合类型主要有 List、Set、Map、Properties，Spring 为这些集合类型提供了专门的配置标签。

1）List 类型的属性

步骤 1．为 Human 类添加一个 List 类型的 favorites 属性。代码如下：

```
public class Human {
......
private List favorites;
public List getFavorites() {
```

```
        return favorites;
    }
    public void setFavorites(List favorites) {
        this.favorites = favorites;
    }
    ......}
```

步骤2．在beans.xml文件中添加favorites属性对应的list配置：read、music、movie。代码如下：

```
<bean id="zhangsan" class="com.Human">
...
<property name="favorites">
<list>
<value>read</value>
<value>music</value>
<value>movie</value>
</list>
</property>
</bean>
```

提示：List类型属性的值既可以是普通值，也可以通过<ref>元素引用其他Bean。

步骤3．修改test方法，专门输出favorites属性值。代码如下：

```
@Test
public void test()
{
    ApplicationContext ac=new ClassPathXmlApplicationContext("applicationContext.
xml");
    Human us=(Human)ac.getBean("zhangsan");
    System.out.println(us.getFavorites());
}
```

步骤4．控制台输出结果如下：

```
[read, music, movie]
```

2）Map类型的属性

步骤1．为Human类添加一个Map类型的relatives属性。代码如下：

```
private Map relatives;
public Map getRelatives() {
    return relatives;
}
public void setRelatives(Map relatives) {
    this.relatives = relatives;
}
```

步骤2．修改配置文件beans.xml，为属性relatives赋予Map类型的值。代码如下：

```
<1,father>,<2,mother>,<3,sister>;清单如下：
<property name="relatives">
<map>
<entry><key><value>1</value></key><value>father</value></entry>
```

```
<entry><key><value>2</value></key><value>mother</value></entry>
<entry><key><value>3</value></key><value>sister</value></entry>
</map>
</property>
```

3）Properties 类型的属性

Properties 类型属性的键和值都只能是字符串。

步骤 1．为 Human 类添加一个 Properties 类型的 pets 属性。代码如下：

```
private Properties pets;
public Properties getPets() {
    return pets;
}
public void setPets(Properties pets) {
    this.pets = pets;
}
```

步骤 2．properties 类型的属性 pets 的配置信息如下：

```
<property name="pets">
<props>
<prop key="dog">hei</prop>
<prop key="cat">tom</prop>
<prop key="pig">maidou</prop>
</props>
</property>
```

6.3.4 Bean 的作用域

定义<bean>时，使用 scope 属性可以指定其作用域。例如，如果需要强制 Spring 每次生成一个新的 Bean 实例，应该将 Bean 的 scope 属性声明为原型。类似地，如果希望 Spring 在每次需要时返回相同 Bean 实例，则应将 Bean 的 scope 属性声明为 singleton（单例模式）。

Spring 支持以下 5 个范围，其中 3 个范围仅在支持 Web 的 ApplicationContext 时才可用。

1）singleton

Bean 定义范围限定为每个 Spring IoC 容器的单个实例（默认）。

2）prototype

Bean 定义范围限定为具有任意数量的对象实例。

3）request

Bean 定义范围限定为 HTTP 请求的生命周期。

4）session

Bean 定义范围限定为 HTTP 会话的生命周期。

5）global session

Bean 定义范围限定为全局 HTTP 会话的生命周期。

如果 Bean 没有定义 scope 属性，则默认为 singleton。如果希望使用原型，要进行如下定义：

```
<bean id="tom" class="com.ssm.util.Car" scope="prototype">
    <property name="type" value="nissan"/>
```

```
        <property name="maxSpeed" value="300"/>
    </bean>
```

6.3.5 延迟初始化 Bean

ApplicationContext 默认会在启动时将所有 singleton Bean 提前实例化。提前实例化是一个良好的习惯，这样配置中或者运行环境中的错误会被立刻发现。如果不希望这样，可以在 singleton Bean 定义时加上延迟加载的属性定义，防止它提前实例化。延迟初始化 Bean 的定义会通知 Spring IoC 容器在第一次需要的时候才实例化，而不是在容器启动时就实例化。

在 XML 配置文件中，延迟初始化通过设置<bean/>元素的 lazy-init 属性为 true 来实现，例如：

```
  <bean id="tom" class="com.ssm.util.Car" scope="prototype" lazy-init="true">
        <property name="type" value="nissan"/>
        <property name="maxSpeed" value="300"/>
    </bean>
```

6.4 基于注解的实例化 Bean

6.4.1 Spring 框架的常用注解

从 Spring 2.0 开始引入了基于注解的配置方式，在 Spring 3.1 中，基于注解的配置方式得到了进一步的完善，使用更为方便。但注解后修改要麻烦一些，耦合度会增加，应该根据需要选择合适的方法。采用 XML 的配置方式时，Bean 的配置信息和实现类是分离的。采用基于注解的配置方式，可以将二者合一。基于注解的配置方式有三个常用的注入注解：@Resource、@Autowired、@Qualifier。

@Autowired 默认按照 Bean 的类型装配注入，如果想按名称来装配注入，则需要结合@Qualifier 一起使用，由@Qualifier 提供 Bean 的名称。

@Resource 默认按照 Bean 的名称来装配注入，只有当找不到与名称匹配的 Bean，才会按照类型来装配注入。

@Resource 由 J2EE 提供，而@Autowired 由 Spring 提供，故为减少系统对 Spring 的依赖，建议使用@Resource 的方式。如果 Maven 项目是在 JRE 1.5 运行的，则需换成更高版本的 Spring。

定义 Bean 时，常用到以下注解：

@Repository：存储的 DAO Bean。

@Service：业务处理 Bean。

@Controller：逻辑控制 Bean。

@Component：上面三种注解不好定义的 Bean，统统注解为 Component（组件）。

6.4.2 任务四：基于注解的实现

【示例 6-4】在示例 6-3 的基础上修改，用基于注解的方式实现 Bean 的定义和属性值注入。

步骤 1．修改配置文件 beans.xml，删掉所有 Bean 的定义，换作以下内容。

```
<context:component-scan base-package="com.ssm.util "/>
```

上面的代码表示 Spring IoC 容器将会自动扫描 com.ssm.util 包下所有的文件。可以对<context: component-scan>属性设置更加精确的范围，如 resource-pattern，对指定的基包下面的子包进行选取。

<context>还有 2 个子标记可以设置更加精准的范围。

include-filter：指定需要包含的包。

exclude-filter：指定需要排除的包。例如：

```
<!-- 自动扫描 com.ssm.util 中的类 -->
<context:component-scan base-package="com.ssm.util" resource-pattern="bo/*.class" />
<context:component-scan base-package="com.ssm" >
  <context:include-filter type="aspectj" expression="com.ssm.dao.*.*"/>
  <context:exclude-filter type="aspectj" expression="com.ssm.entity.*.*"/>
</context:component-scan>
```

include-filter 表示需要包含的目标类型，exclude-filter 表示需要排除的目标类型，type 表示采用的过滤类型，共有 5 种，如 annotation、aspectj 等，详细可以查看 Spring 官方文档。expression 表示过滤的表达式。

步骤 2．修改 com.ssm.util/Car.java 文件，给类添加@Component("myBmw")的注解，生成一个名为 myBmw 的 Bean。并且，给属性 type 添加 @Value("BMW")的注解，表示给 type 赋值 BMW；给属性 maxSpeed 添加@Value("500")的注解，表示给 maxSpeed 赋值 500。代码如下：

```
@Component("myBmw")
public class Car {
    @Value("BMW")
    private String type;
    @Value("500")
    private int maxSpeed;
    public Car() {   }
    public Car(String type, int maxSpeed) {
        this.type = type;
        this.maxSpeed = maxSpeed;   }
    public String getType() {
        return type;    }
    public void setType(String type) {
        this.type = type;    }
    public int getMaxSpeed() {
        return maxSpeed;    }
    public void setMaxSpeed(int maxSpeed) {
        this.maxSpeed = maxSpeed;    }
    public void introduce(){
        System.out.println("type is: "+ getType()+" ,maxSpeed is: "+getMaxSpeed ());
    }
}
```

步骤 3. 修改 com.ssm.util/Human.java 文件，给 Human 类添加注解@Component("zhangsan")，定义一个名称为 zhangsan 的 Bean；并且给属性 name 添加注解@Value("张三")，赋值为张三，给属性 car 添加注解@Autowired，表示这里存在一个自动注入的关系，@Autowired 后面紧随@Qualifier("myBmw")，表示注入的对象是前面定义过的 Bean，按名称 myBmw 查找该 Bean。代码如下：

```
@Component("zhangsan")
public class Human {
    @Value("张三")
    private String name;
    @Autowired
    @Qualifier("myBmw")
    private Car car;
    public String getName() {
        return name;    }
    public void setName(String name) {
        this.name = name;    }
    public Car getCar() {
        return car;    }
    public void setCar(Car car) {
        this.car = car;    }
    public void introduce(){
        System.out.println(getName()+" , car:");
        car.introduce();    }
}
```

步骤 4. 和前面的示例 6-3 一样运行 test 方法，控制台输出如下结果：

```
张三 , car:
type is: BMW ,maxSpeed is: 500
```

在示例 6-4 中我们看到，普通属性值的注入使用注解@Value，引用对象的注入则使用注解@Autowired 和@Qualifier。

6.5 IoC 的零配置实现

零配置就是 Spring 不再使用 beans.xml 文件来初始化容器，而使用一个类来替代。

【示例 6-5】使用零配置来实现示例 6-4。

步骤 1. 新增一个用于替代原 XML 配置文件的 ApplicationCfg 类。代码如下：

```
package com.ssm.util;
import org.springframework.context.annotation.Bean;
import org.springframework.context.annotation.ComponentScan;
import org.springframework.context.annotation.Configuration;
@Configuration
@ComponentScan(basePackages="com.ssm.util")
```

```
public class ApplicationCfg {
    @Bean
    public Human getHuman(){
        return new Human();
    }
}
```

@Configuration 相当于配置文件中的<beans/>，@ComponentScan 相当于配置文件中的 context:component-scan，属性 basePackages 也一样设置为 com.ssm.util 包。@Bean 相当于<bean/>，只能注解在方法和注解上，一般在方法上使用，源码中描述：@Target({ElementType.METHOD, ElementType.ANNOTATION_TYPE})，方法名相当于 id。中间用到 Human 类，Human 类的代码无需修改。

步骤 2．初始化容器 Test 类的代码与以前略有不同，具体如下：

```
package com.ssm.util;
import org.springframework.context.ApplicationContext;
import org.springframework.context.annotation.AnnotationConfigApplicationContext;
public class Test {
    @org.junit.Test
    public void test(){
        ApplicationContext applicationContext=new AnnotationConfigApplication Context
        (ApplicationCfg.class);
        Human zhangsan=(Human)applicationContext.getBean("zhangsan");
        zhangsan.introduce();
    }
}
```

容器的初始化通过注解 AnnotationConfigApplicationContext 类完成，即用 ApplicationCfg.class 初始化容器。中间 zhangsan 与 getHuman 不是相同的 Bean，因为在 ApplicationCfg 中声明的方法 getBean 相当于在 XML 文件中定义了一个<bean id="getHuman" class="..."/>，在 Human 类上注解@Component("zhangsan")相当于另一个<bean id="zhangsan" class="..."/>。

步骤 3．运行 test 方法，控制台输出如下结果：

```
    2019-11-04-10-57 [main] [org.springframework.context.annotation.AnnotationConfig
ApplicationContext] [INFO] - Refreshing org.springframework.context.annotation.
AnnotationConfigApplicationContext@591f989e: startup date [Mon Nov 04 10:57:55 CST 2019];
root of context hierarchy
    张三 , car:
    type is: BMW ,maxSpeed is: 500
    Process finished with exit code 0
```

从输出结果看，zhangsan 的属性 name 和 car 都已成功注入。

小结：使用零配置注解虽然方便，不需要编写复杂的 XML 文件，但并非可以完全取代 XML 文件。应该根据实际需要选择，或二者结合使用。毕竟使用类作为容器的配置信息是硬编码，发布后不容易修改。

6.6 项目中 Spring IoC 的使用

在我们一直使用的、集成了各大框架的示例 StudentGradle 项目中，早已使用了 Spring IoC 容器来创建和管理各种 Bean。下面从 3 个方面来讲解其在 StudentGradle 中的使用。

6.6.1 WebApplicationContext

WebApplicationContext 是 ApplicaitonContext 的实现类，是专门为 Web 应用准备的。它允许从相对于 Web 根目录的路径中装载配置文件，完成 Bean 的实例化的初始工作。从 WebApplicationContext 中可以获得 ServletContext 的引用，整个 Web 应用上下文对象将作为属性放置在 ServletContext 中，以便 Web 应用可以访问 Spring 上下文。Spring 中提供 WebApplicationContextUtils 的 getWebApplicationContext(ServletContext src) 方法来获得 WebApplicationContext 对象。

有过 Java Web 开发经验的读者都知道在 web.xml 文件中配置 Servlet 或者定义 Web 容器监听器，就可以完成 Spring Web 应用上下文的启动。下面截取项目中 web.xml 的部分配置信息来讲解：

```
    <!--从类路径下加载 Spring 配置文件，classpath 特指类路径下加载-->
    <context-param>
        <param-name>contextConfigLocation</param-name>
        <param-value>classpath*:spring-mybatis.xml</param-value>
    </context-param>
    <!-- 配置 Spring 的监听器 -->
    <listener>
    <listener-class>org.springframework.web.context.ContextLoaderListener
</listener-class>
    </listener>
```

ContextLoaderListener 通过 Web 容器上下文参数 contextConfigLocation 获取 Spring 配置文件，即位于类路径下的 spring-mybatis.xml。用户可以指定多个配置文件，中间用逗号隔开即可。

6.6.2 项目使用 XML 配置的场景

无论在 spring-mvc.xml 或者 spring-mybatis.xml 中都有大量 Bean，这些 Bean 都是用 Spring 或者 MyBatis 库的类创建的。

例如，在 spring-mvc.xml 中有这样的 Bean 配置：

```
  <!-- 视图渲染 -->
    <bean id="internalResourceViewResolver" class="org.Springframework.web.servlet.
view.InternalResourceViewResolver">
        <property name="prefix" value="/WEB-INF/views/"/>
        <property name="suffix" value=".jsp"/>
```

```
    </bean>
```

这个 Bean 定义了内部的视图解析器，它由 org.Springframework.web.servlet.view.InternalResourceViewResolver 类创建，并且定义了所有的视图都是 JSP 页面，存放于/WEB-INF/views/目录。

而在 spring-mybatis.xml 中有这样的 Bean 定义：

```
  <bean id="dataSource" class="com.mchange.v2.c3p0.ComboPooledDataSource" destroy-
method="close">
          <property name="driverClass" value="${jdbc.driver}"/>
          <property name="jdbcUrl" value="${jdbc.url}"/>
          <property name="user" value="${jdbc.username}"/>
          <property name="password" value="${jdbc.password}"/>
  </bean>
```

上面的配置信息定义了一个名称为 dataSource 的 Bean，使用 C3P0 的数据库连接池类创建，并提供了 4 个连接属性值：驱动器类 driverClass、连接字符串 jdbcUrl、数据库用户名 user、数据库连接密码 password。destroy-method="close"表示 Bean 销毁时调用 close 方法。

紧随 dataSource 定义后，另一个 Bean 的定义如下：

```
  <!-- spring 和 Mybatis 完美整合，不需要 Mybatis 的配置映射文件 -->
      <bean id="sqlSessionFactory" class="org.mybatis.spring.SqlSessionFactoryBean">
          <property name="dataSource" ref="dataSource" />
          <!-- 自动扫描 mapping.xml 文件 -->
          <property name="mapperLocations" value="classpath*:/mapper/*.xml"></property>
          <!-- 配置分页插件 -->
         ......
          </property>
      </bean>
```

sqlSessionFactory 由 org.mybatis.spring.SqlSessionFactoryBean 实例化，它里面有个属性 dataSource 引用了上面的 Bean（dataSource），并且定义了数据库映射文件所在的位置，即类路径下 mapper/*.xml。

6.6.3 项目使用注解配置的场景

基于注解的配置在我们的项目中也被广泛使用。例如，在 spring-mvc.xml 文件中，使用 xmlns:context="http://www.Springframework.org/schema/context"声明 context 的命名空间，然后使用 context 命名空间下的 component-scan 标签的属性 base-package 来指定需要扫描的基类包。<context:component-scan base-package="com.ssm"/>定义了自动扫描包的范围是 com.ssm。

在 spring-mybatis.xml 文件中也定义了一个自动扫描。com.ssm.dao 是 DAO 接口所在包的包名，Spring 会自动查找其下的类。代码如下：

```
      <bean class="org.Mybatis.Spring.mapper.MapperScannerConfigurer">
        <property name="basePackage" value="com.ssm.dao" />
        <property name="sqlSessionFactoryBeanName"   value="sqlSessionFactory">
          </property>
  </bean>
```

项目中，controller 自动注入 Service 类型的 Bean，在 Service 类型的实现类里又自动注入 DAO 类型的 Bean。例如，在 com.ssm.controller.StudentController 类型里，我们使用@Controller 为 StudentController 类创建了一个名为 studentController 的 Bean。使用下面的代码为 student Controller 自动注入名为 studentServiceImpl 的 Bean。

```
@Autowired
   private StudentService studentServiceImpl;
```

这个 Bean 在哪儿创建的？我们打开 com.ssm.service.impl.StudentServiceImpl 文件，StudentService 类的上面有这样一个注解@Service("studentServiceImpl")，它定义了一个名为 studentServiceImpl 的 Service 类型的 bean。在 StudentServiceImpl 类里，它又通过使用下面的代码自动注入了名为 studentMapper 的 DAO 类型的 Bean。studentMapper 是 Spring 容器使用 MapperScannerConfigurer 类扫描时创建的 Bean。

```
@Autowired
private StudentMapper studentMapper;
```

提示：如果 StudentService 类上的注解@Service 没有定义 Bean 的名称，则使用默认的命名机制，将 Bean 的名称默认为把类的首字母小写，其他字母不变。例如，StudentServiceImpl 类的 Bean 的名则是 studentServiceImpl，TeacherServiceImpl 类的 Bean 的名则为 teacherServiceImpl。

下面我们以 StudentServiceImpl 为例说明其注入过程。

其实这里我们只需要一个类，但在实际项目开发的过程中，一般会使用面向接口编程。所以这里使用接口与接口实现的形式。代码如下：

```
StudentService.java 的代码清单如下：
public interface StudentService {
   List<Student>  listStudents();                      //显示所有学生
   void insert(Student student);                       //插入新学生
   void delete(int id);                                //通过 id 号删除学生
   Student find(int id);                               //通过 id 号查询学生
   void update(Student student);                       //更新学生信息
   Student login(String name,String password);         //登录
   //获取分页查询学生信息
   List<Student> findByPaging(Integer toPageNo);
}
```

StudentService 是一个接口，里面定义了很多数据库操作的业务方法。

接下来，我们给出实现类 StudentServiceImpl.java 的代码，里面使用了@Service ("studentService")定义 Service 类型的 Bean，然后使用@Autowired、private StudentMapper studentMapper 自动注入 DAO 类型的 Bean。

```
@Service("studentService")
public class StudentServiceImpl implements StudentService{
   @Autowired
   private StudentMapper studentMapper;
   @Override
   public List<Student> listStudents() {
      return studentMapper.listStudents();
   }
```

```
    @Override
    public void insert(Student student) {
        studentMapper.insert(student);
    }
    @Override
    public void delete(int id) {
        studentMapper.deleteByPrimaryKey(id);
    }
    @Override
    public Student find(int id) {
        return studentMapper.selectByPrimaryKey(id);
    }
    @Override
    public void update(Student student) {
        studentMapper.updateByPrimaryKey(student);
    }
    @Override
    public Student login(String name, String password) {
        return studentMapper.login(name ,password);
    }
        @Override
    public List<Student> findByPaging(Integer toPageNo) {
        Page page = new Page();
        page.setToPageNo(toPageNo);
        List<Student> list = studentMapper.findByPaging(page);
        return list;
    }
}
```

Service 类型 Bean 的创建方法是 Spring IoC 容器读取 spring-mvc.xml 文件后，读取<context:component-scan base-package="com.ssm"/>并扫描 com.ssm 包下所有的类。

6.7 拓展知识：通过静态工厂方法和实例工厂方法获取 Bean

在 6.3 节中，我们使用构造器通过类名在配置文件中配置 Bean，然后实例化 Bean。事实上还可以通过其他设计模式如静态工厂方法实例化 Bean。下面还是在示例 6-3 的基础上修改，介绍通过静态工厂方法和实例工厂方法获取 Bean。

6.7.1 任务五：用静态工厂方法获取 Bean

【示例 6-6】用静态工厂方法获取 Bean。

步骤 1．创建 Car 的静态工厂类，定义 Map 类型的属性，存放生成的两个 Bean：nissan 和 bmw；定义类的静态方法 getCar，通过 name 返回 Bean。代码如下：

```
package com.ssm.util;
import java.util.HashMap;
```

```
import java.util.Map;
public class StaticCarFactory {
    private static Map<String ,Car> cars=new HashMap<>();
    static {
        cars.put("nissan",new Car("nissan",300));
        cars.put("bmw",new Car("bmw",500));
    }
    public static Car getCar(String name){
        return cars.get(name);
    }
}
```

步骤 2. 修改 beans.xml 文件，Bean 的 class 属性指向静态工厂方法的全类名 StaticCarFactory；factory-method 属性指向静态工厂方法的名称 getCar；<constructor-arg>标签表示如果工厂方法需要传入参数，则使用<constructor-arg>来配置参数。代码如下：

```
<?xml version="1.0" encoding="UTF-8"?>
<beans xsi:schemaLocation="http://www.springframework.org/schema/beans http://www.springframework.org/schema/beans/spring-beans-3.1.xsd    http://www.springframework.org/schema/context http://www.springframework.org/schema/context/spring-context.xsd"
       xmlns:p="http://www.springframework.org/schema/p"
       xmlns:xsi="http://www.w3.org/2001/XMLSchema-instance"
       xmlns="http://www. springframework.org/schema/beans"
       xmlns:context="http://www.springframework.org/schema/context">
    <bean id="nissan" class="com.ssm.util.StaticCarFactory" factory-method= "getCar">
        <constructor-arg value="nissan"/>
    </bean>
    <bean id="bmw" class="com.ssm.util.StaticCarFactory"  factory-method="getCar">
        <constructor-arg value="bmw"/>
    </bean>
    <!--<context:component-scan base-package="com.ssm.util "/>-->
</beans>
```

步骤 3．修改 Test.java 文件，使用 ClassPathXmlApplicationContext 获取两个 Bean：nissan 和 bmw。代码如下：

```
package com.ssm.util;
import org.springframework.context.ApplicationContext;
import org.springframework.context.annotation.AnnotationConfigApplicationContext;
import org.springframework.context.support.ClassPathXmlApplicationContext;
public class Test {
    @org.junit.Test
    public void test(){
        ApplicationContext  applicationContext=new  ClassPathXmlApplicationContext
("beans.xml");
        Car nissan= (Car) applicationContext.getBean("nissan");
        nissan.introduce();
        Car bmw= (Car) applicationContext.getBean("bmw");
        bmw.introduce();
```

```
    }
}
```

步骤 4．运行 test 方法，控制台输出如下信息：

```
type is: nissan ,maxSpeed is: 300
type is: bmw ,maxSpeed is: 500
```

6.7.2　任务六：用实例工厂方法获取 Bean

用实例工厂方法获取 Bean 需要先创建工厂本身，再调用工厂的实例方法返回 Bean 的实例。

【示例 6-7】用实例工厂方法获取 Bean。

步骤 1．创建实例工厂，工厂有个空构造方法，其中 Map 类型的 cars 里存放生成的两个 Bean：nissan 和 bmw。代码如下：

```
package com.ssm.util;
import java.util.HashMap;
import java.util.Map;
public class InstanceCarFactory {
    private Map<String, Car> cars = null;
    public InstanceCarFactory() {
        cars = new HashMap<>();
        cars.put("nissan", new Car("nissan", 500));
        cars.put("bmw", new Car("bmw", 300));
    }
    public Car getCar(String name) {
        return cars.get(name);
    }
}
```

步骤 2．修改 beans.xml 文件，先使用 InstanceCarFactory 类创建工厂 carFactory，然后将 Bean 的 factory-bean 属性指向实例工厂 carFactory，factory-method 属性指向静态工厂方法的名称 getCar。<constructor-arg>标签用来配置参数。代码如下：

```
<?xml version="1.0" encoding="UTF-8"?>
<beans xsi:schemaLocation="http://www.springframework.org/schema/beans http://www.
springframework.org/schema/beans/spring-beans-3.1.xsd http://www.springframework.org/
schema/context http://www.springframework.org/schema/context/spring-context.xsd"
        xmlns:p="http://www.springframework.org/schema/p"
        xmlns:xsi="http://www.w3.org/2001/XMLSchema-instance"
        xmlns="http://www. springframework.org/schema/beans"
        xmlns:context="http://www.springframework.org/schema/context">
    <bean id="carFactory" class="com.ssm.util.InstanceCarFactory"/>
    <bean id="nissan" factory-bean="carFactory" factory-method="getCar">
        <constructor-arg value="nissan"/>
    </bean>
    <bean id="bmw" factory-bean="carFactory" factory-method="getCar">
        <constructor-arg value="bmw"/>
    </bean>
</beans>
```

步骤 3．test 方法无须修改，直接运行，控制台仍然输出如下信息：

```
type is: nissan ,maxSpeed is: 500
type is: bmw ,maxSpeed is: 300
```

小　结

Spring 框架作为一个免费开源的轻量级框架，极大地简化了复杂项目的构建过程。它对代码是非侵入式的，通过 IoC 原理降低了组件间的耦合度，实现了软件各层的解耦。同时，它的模块化特征使得 Spring 发展极为迅速。无论在 Web Framework、Cloud 还是安全认证、工作流等领域，Spring 都能独树一帜。

本章通过丰富的案例着重讲解 Spring IoC 的原理和实现，以及在各种场景中的配置和使用步骤。本章内容与我们原项目编码貌似没有很大关系，但在项目的实际运作中起着至关重要的作用。请大家细细体会。

习　题

一、单选题

1．下面属于 IoC 自动装载方法的是（　　）。

A．byMethod　　B．method　　C．constructor　　D．byName

2．下面有关 Spring 依赖注入的说法中正确的是（　　）。

A．IoC 就是由 Spring 负责控制对象的生命周期和对象间的关系的

B．BeanFactory 是最简单的容器，提供了基础的依赖注入支持

C．ApplicationContext 建立在 BeanFactory 之上，提供了系统构架服务

D．如果 Bean 的某一个属性没有注入，ApplicationContext 加载后，直至第一次调用 getBean 方法才会抛出异常；而 BeanFactory 则在初始化自身时检验，这样有利于检查所依赖属性是否注入

3．下列关于 Spring 特性中 IoC 的描述中正确的是（　　）。

A．IoC 就是指程序之间的关系由程序代码直接操控

B．“控制反转”是指控制权由应用代码转到外部容器，即控制权的转移

C．IoC 将控制创建的职责搬进了框架中，从应用代码脱离开来

D．使用 Spring 的 IoC 容器时只需指出组件需要的对象，在运行时 Spring 的 IoC 容器会根据 XML 配置数据提供给它

二、填空题

1．________是 Spring 容器的内核，AOP、声明式事务都是在此基础上开花结果的。

2．IoC 思想的基础是________，组件之间的依赖关系由容器在运行期决定，即由容器动态地将某个依赖关系注入组件之中。

3．Spring IoC 基于注解的配置里有三个常用的注入注解________、________、________。

4．________是 ApplicaitonContext 的实现类， 是专门为 Web 应用准备的，它允许从相对于 web 根目录的路径中装载配置文件完成 Bean 的实例化的初始工作。

三、简答题

1．Spring IoC 是如何使用 Java 反射机制实现控制反转的？

2．Spring IoC 容器是如何管理 Bean 的？

综合实训

实训 1．创建一头牛类，属性：名称、体重。创建一辆牛车类，属性：牛、车轮个数。使用 XML 文件配置上面两个类。牛：niumowang，500。牛车：niumowang，3 个轮子。使用 JUnit 4 测试，输出牛车的信息。

实训 2．使用注解配置实训 1 内容。

第7章

项目集成 Spring AOP

本章学习目标

- 了解 Spring AOP 的原理
- 掌握 Spring AOP 基于 XML 和注解的实现
- 掌握 Spring 事务管理配置

本章介绍了 AOP 及 Spring AOP 的实现、事务管理等基础知识，通过案例详细阐述了如何使用 Spring AOP 实现各种功能。最后，使用 Spring IoC 和 AOP 将整个项目集成并运行起来。

7.1 AOP

AOP（Aspect Oriented Programming，面向切面编程）是一个概念，也是一个规范，本身并没有设定具体语言的实现，但实际上提供了非常广阔的发展空间。这种编程思想抽取出散落在软件系统各处的横切关注点代码，并将其模块化，然后归整到一起，从而使得业务逻辑各部分之间的耦合度降低，进一步提高软件的可维护性、复用性和可扩展性。

它主要使用的场景为日志记录、性能统计、安全控制、事务处理、异常处理等。可以将日志记录、性能统计、安全控制、事务处理、异常处理等代码从业务逻辑代码中分离出来，并将它们放入非指导业务逻辑的方法中，实现改变这些行为的时候不影响业务逻辑的代码。

7.1.1 AOP 概述

传统的面向对象编程中，每个单元就是一个类，而类似于安全性的问题通常不能集中在一个类中处理，因为安全问题横跨多个类，遍布整个项目。面向对象编程（OOP）面临的项目状态如图 7-1 所示。

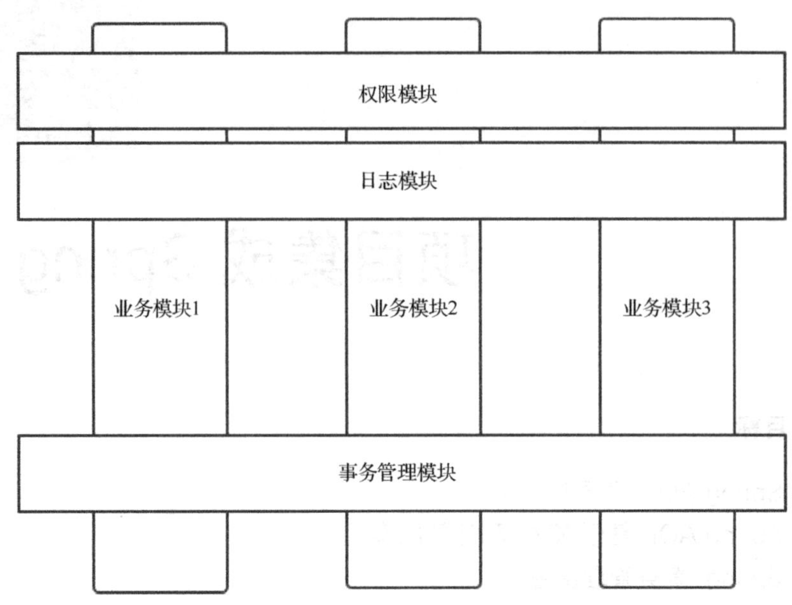

图 7-1　OOP 面临的项目状态

这样的编码方式导致代码无法重用。OOP 实现起来就如下面的业务代码：

```
public Teacher selectCoursesById(int id) {
    transactionManager.beginTransaction();
    return teacherMapper.selectCourseById(id);
    transactionManager.commit();
}
```

我们会发现，业务代码被重复的非业务代码包围，让开发者不能集中精力在业务逻辑处理上，并且上面的代码也不便于阅读。

AOP 应运而生，专门解决一些具有横切性质的系统性服务，如事务管理、安全检查、缓存管理、对象池管理等。可以这样理解，OOP 是从静态角度考虑程序结构的，AOP 是从动态角度考虑程序运行过程的。AOP 是 OOP 的延续，是对 OOP 的补充。AOP 通过横向切割的方式抽取分散在业务代码中的相同代码，构成一个独立的模块，还业务逻辑类一个清晰的世界。

AOP 可以分为静态织入与动态织入。静态织入即在编译前将需织入内容写入目标模块中，这样成本非常高。动态织入则不需要改变目标模块。AOP 通过预编译方式和运行期动态代理，实现了在不修改源代码的情况下给程序动态、统一添加功能。AOP 实际是 GoF 设计模式的延续，追求的是调用者和被调用者之间的解耦，以提高代码的灵活性和可扩展性。

那么，如何将抽取出来的非业务逻辑代码再次还原到原代码中呢？这正是 AOP 要致力解决的问题，也是本章接下来的教学重点。

7.1.2　AOP 术语

1. 连接点（Join Point）

在一段代码中，具有边界性质的特定点被称为连接点。Spring 仅支持方法的连接点，即在方法调用前、方法调用后、方法抛出异常时为这些程序执行点织入增强。AOP 向目标类切入时的候选点被称为连接点。

连接点由两个信息确定：一是用方法表示的程序执行的位置；二是用相对点表示的方位（前、后、环绕等）。如在 UserDAOImpl.addUser() 方法执行前的连接点，执行点为 UserDAOImpl.addUser()，方位为该方法执行前的位置。Spring 使用切点对执行点进行定位，而方位则在增强类型中定义。

2. 切点（Pointcut）

一个拥有多个方法的类，会拥有多个连接点。在为数众多的连接点中，AOP 通过切点定位特定的连接点。切点和连接点不是一对一的关系，一个切点可以匹配多个连接点。

在 Spring 中，切点通过 org.Springframework.aop.Pointcut 接口进行描述，它使用类和方法作为连接点的查询条件。Spring AOP 的规则解析引擎负责解析切点所设定的查询条件，找到对应的连接点。确切地说，应该是执行点而非连接点，因为连接点是方法执行前、执行后等包含方位信息的具体程序执行点，而切点只定位到某个方法上，所以说如果希望定位到某个连接点上，还需要提供方位信息。

3. 增强（Advice）

增强，有些文献也翻译成通知，它是织入目标类连接点上的一段程序代码。在 Spring 中，增强除了用于描述一段程序代码，还拥有另一个和连接点相关的信息，这便是执行点的方位。结合执行点方位信息和切点信息，我们就可以找到特定的连接点。正因为增强既包含了用于添加到目标连接点上的一段执行逻辑，也包含用于定位连接点的方位信息，所以 Spring 所提供的增强接口都是带方位名的，如前置增强 BeforeAdvice。所以只有将切点和增强两者结合，才能确定特定的连接点并实施增强逻辑。

4. 目标对象（Target）

增强逻辑的织入目标类。如果没有 AOP，目标业务类需要自己实现所有逻辑，如图 7-1 所示。在 AOP 的帮助下，业务模块只需要实现那些非横切逻辑的程序逻辑，而性能监视和事务管理等这些横切逻辑则可以使用 AOP 动态织入特定的连接点。

5. 织入（Weaving）

织入是将增强添加到目标类具体连接点上的过程，AOP 像一台织布机，将目标类和增强天衣无缝地编织在一起。AOP 有以下 3 种织入方式：

（1）编译期织入。

（2）类装载期织入。

（3）动态代理织入。

6. 代理（Proxy）

一个类被 AOP 织入增强后，就产生了一个新的结果类，它是将目标类和增强逻辑编织在一起的代理类。根据不同的代理方式，代理类可能是和原类具有相同的接口的类，也可能是原类的子类，所以可以采用与调用原类相同的方式调用代理类。

7. 切面（Aspect）

切面由切点和增强组成，它既包括横切逻辑的定义，也包括连接点的定义。Spring AOP 就是负责实施切面的框架，它将切面所定义的横切逻辑织入切面所指定的连接点中。

7.2 Spring AOP

Spring 中 AOP 代理由 Spring 的 IoC 容器负责生成、管理，其依赖关系也由 IoC 容器负责管理。因此，AOP 代理可以直接使用容器中的其他 Bean 实例作为目标，这种关系可由 IoC 容器的依赖注入提供。Spring 默认使用 Java 动态代理来创建 AOP 代理，这样就可以为任何接口实例创建代理了。当需要代理的类不是代理接口的时候，Spring 会自动切换为使用 CGLIB 代理，也可强制使用 CGLIB。

AOP 编程其实是很简单的事情，纵观 AOP 编程，程序员只需要参与三个部分：

（1）定义普通业务组件。

（2）定义切点，一个切点可能横切多个业务组件。

（3）定义增强处理，增强处理就是在 AOP 框架为普通业务组件织入的处理动作。

所以进行 AOP 编程的关键就是定义切点和定义增强处理，一旦定义了合适的切点和增强处理，AOP 框架将自动生成 AOP 代理，即代理对象的方法=增强处理+被代理对象的方法。

Spring AOP 是 AOP 技术在 Spring 中的具体实现，它与 IoC 一起构成了 Spring 的重要技术。Spring 3.1 增加了对 AspectJ 更直接的支持。

7.2.1 AspectJ

作为 AOP 的具体实现之一的 AspectJ，是一个完整的面向切面编程的框架。它扩展了 Java 语言，JDK1.4 版本之后就支持 AspectJ。AspectJ 定义了 AOP 的语法，它有一个专门的编译器用来生成遵守 Java 字节编码规范的 Class 文件。AspectJ 还支持原生的 Java，只需要加上 AspectJ 提供的注解即可。它给 Java 中加入了连接点这个新概念，还向 Java 语言中加入了一些新成分，如切点、通知、类型间声明（Inter-type Declaration）和切面。Spring AOP 沿用了 AspectJ 中定义的 AOP 方面的术语。

7.2.2 Spring AOP 与 AspectJ 的关系

AspectJ 是一个完整的面向切面编程的实现框架，而 Spring AOP 的目的并不是提供最完整的 AOP 实现，而是帮助解决企业应用中的常见问题，实现一个 AOP 与 Spring IoC 之间的紧密集成。并且，Spring AOP 非常容易实现，可以在 Spring Bean 之上将横切关注点模块化。

Spring AOP 致力于提供一种能够与 Spring IoC 紧密集成的面向切面框架的实现，以便于解决在开发企业级项目时面临的常见问题。在选择使用 Spring AOP 或者 AspectJ 前，需要明确应用的横切关注点（如事务管理、日志或性能评估），需要处理的是 Spring Bean 还是 POJO。如果是开发新应用，则选择 Spring AOP。如果是维护一个已有的应用（该应用并没有使用 Spring 框架），AspectJ 就将是一个不错的选择。

另一个需要考虑的因素是，你是希望在编译期间织入，还是编译后或是运行时织入。Spring 只支持运行时织入，如果期望在编译期间织入，只能选择 AspectJ。

7.2.3 Spring AOP 增强

AOP 的工作重心在于如何将增强应用于目标对象的连接点上，这里包括两个工作：第一，如何通过切点和增强定位到连接点上；第二，如何在增强中编写切面的代码。

按增强在目标类中方法的连接点位置，Spring 支持 5 种类型的增强：

前置增强：org.Springframework.aop.MethodBeforeAdvice，表示在目标方法执行前实施增强。

后置增强：org.Springframework.aop.AfterReturningAdvice，表示在目标方法执行后实施增强。

环绕增强：org.aopalliance.intercept.MethodInterceptor，表示在目标方法执行前后实施增强。

异常抛出增强：org.Springframework.aop.ThrowsAdvice，表示在目标方法抛出异常后实施增强。

引介增强：org.Springframework.aop.support.DelegatingIntroductionInterceptor，表示在目标类中添加一些新的方法和属性。

引介增强是一种特殊的增强，它为类添加一些属性和方法。这样，即使一个业务类原本没有实现某个接口，通过 AOP 的引介功能，也可以动态地为该事务添加接口的实现逻辑，让业务类成为这个接口的实现类。

7.2.4 添加 Spring AOP 依赖

在 Web 项目中使用 Spring AOP 之前，需要添加对它的依赖。因为我们之前在 build.gradle 文件中引入了 spring-web 依赖，里面包含了 Spring AOP 及其需要的其他包。

如果希望使用 AspectJ 的注解或类，需要在 build.gradle 文件添加对 AspectJ 的依赖。

```
// https://mvnrepository.com/artifact/org.aspectj/aspectjweaver
compile group: 'org.aspectj', name: 'aspectjweaver', version: '1.9.2'
```

具体依赖如图 7-2 所示。

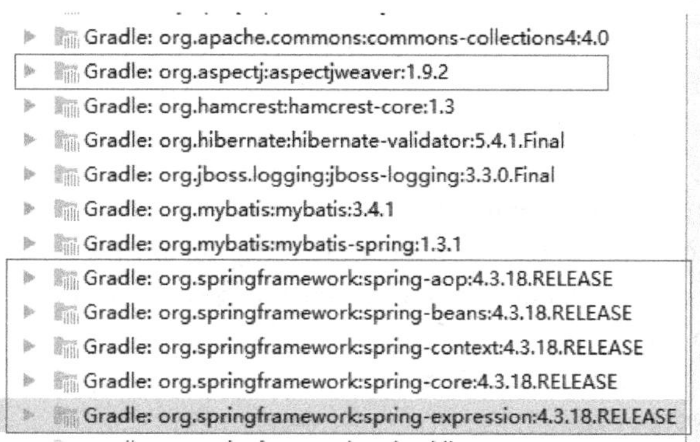

图 7-2 Spring AOP 和 AspectJ 依赖

7.2.5 任务一：动态代理实现之 JDK 动态代理

为了更好地理解 Spring 的 AOP 技术，我们先手动编写两种使用反射机制实现动态代理的方法，从中体会用 Spring 实现 AOP 技术所带来的便利。

【示例 7-1】JDK 动态代理。

提示：在实现 JDK 动态代理的过程中，目标代理接口和目标代理类都需要被定义。

步骤 1．借用 TeacherService 和 TeacherServiceImpl 作为目标代理接口和目标代理类。添加一个 print 方法（用于输出字符），作为调用的业务方法，即 AOP 中的连接点。代码如下：

```
public interface TeacherService {
    ...
    public void print();
}
public class TeacherServiceImpl implements TeacherService {
    ...
    @Override
    public void print() {
        System.out.println("teacher list");
    }
}
```

步骤 2．创建切面 MyAspect，里面有两个方法：before 方法将作为前置方法，after 方法将作为后置方法。代码如下：

```
public class MyAspect {
    public void before(){
        System.out.println("before advice");
    }
    public void after(){
        System.out.println("after advice");
}}
```

步骤 3．创建代理工厂 ProxyFactory，该工厂创建被代理类的代理类对象，通过加载该被代理类的类和接口，按顺序执行增强和即将执行的业务方法。代码如下：

```
package com.ssm.util;
import java.lang.reflect.InvocationHandler;
import java.lang.reflect.Method;
import java.lang.reflect.Proxy;
public class ProxyFactory {
    final static MyAspect ma=new MyAspect();
    public static Object CreateProxy(final Object us){
        Object obj= Proxy.newProxyInstance(us.getClass().getClassLoader(),us.get
Class().getInterfaces(),
                new InvocationHandler() {
                    @Override
                    public Object invoke(Object proxy, Method method, Object[] args)
throws Throwable {
                        ma.before();
                        Object obj1=method.invoke(us,args);
                        ma.after();
                       return obj1;
                    }
```

```
            };
        return obj;
    }
}
```

步骤 4. 创建测试方法，使用 TeacherServiceImpl 类新建被代理对象 ts，使用代理工厂 Proxy Factory 创建代理对象 proxyTeacherService。当代理对象调用 print 方法时，调用代理对象的回调方法，实现增强的切入。代码如下：

```
public class Test {
    @org.junit.Test
    public void test(){
        TeacherService ts=new TeacherServiceImpl();
         TeacherService proxyTeacherService= (TeacherService) ProxyFactory. Create
Proxy(ts);
        proxyTeacherService.print();
    }
}
```

步骤 5. 运行测试方法 Test.test，控制台输出如下信息，即增强的 before 方法运行在 print 方法之前，after 方法运行在 print 方法之后。代码如下：

```
before advice
teacher list
after advice
```

7.2.6 任务二：动态代理实现之 CGLIB 字节码增强

CGLIB（Code Generator Library）是一个强大的、高性能的代码生成库，CGLIB 采用非常底层的字节码技术。其被广泛应用于 AOP 框架（Spring、dynaop）中，用以提供方法拦截操作。采用 CGLIB 实现动态代理时，被代理的对象不需要实现接口，很多时候都采用 CGLIB 来进行动态代理。Spring 已经将 CGLIB 和 ASM 整合到 spring-core-3.2.0.jar 中，所以导入 Spring 的 jar 包后就不需要再重复导入这两个包了。

【示例 7-2】CGLIB 动态代理。

步骤 1. 创建不需要实现接口的被代理类 TeacherServiceImpl，定义其 print 方法。代码如下：

```
public class TeacherServiceImpl  {
       public void print() {
        System.out.println("teacher list");
    }
}
```

步骤 2. 切面 MyAspect 不变，里面有两个方法：before 方法将作为前置方法，after 方法将作为后置方法。代码如下：

```
public class MyAspect {
    public void before(){
        System.out.println("before advice");
    }
    public void after(){
```

```
        System.out.println("after advice");
}}
```

步骤 3．创建代理工厂 ProxyFactory ，该工厂创建被代理类的代理类对象，通过加载该被代理类的类和接口，按顺序将前置增强和后置增强分别织入业务方法。业务方法被 invoke 和 invokeSuper 执行了两次。代码如下：

```
public class ProxyFactory {
    final static  TeacherServiceImpl teacherService=new TeacherServiceImpl();
    final static MyAspect ma=new MyAspect();
    public static TeacherServiceImpl CreateProxy(){
        Enhancer enhancer=new Enhancer();
        enhancer.setSuperclass(teacherService.getClass();
        enhancer.setCallback(new MethodInterceptor() {
            @Override
            public Object  intercept(Object  proxy,  Method  method,  Object[]  args, 
MethodProxy methodProxy) throws Throwable {
                  ma.before();
                  Object obj=method.invoke(teacherService,args);
                  methodProxy.invokeSuper(proxy,args);
                  ma.after();
                  return  obj;
            }
        };
        return  (TeacherServiceImpl)enhancer.create();
    }
}
```

步骤 4．创建测试方法，使用 TeacherServiceImpl 类新建被代理对象 ts，使用代理工厂 ProxyFactory 创建代理对象 proxyTeacherService。当代理对象调用 print 方法时，调用代理对象的回调方法，实现增强的切入。代码如下：

```
public class Test {
    @org.junit.Test
    public void test(){
        TeacherServiceImpl ts=new TeacherServiceImpl();
        TeacherServiceImpl proxyTeacherService= (TeacherServiceImpl) ProxyFactory.
CreateProxy();
        proxyTeacherService.print();
    }
}
```

步骤 5．运行测试方法 Test.test，控制台输出如下信息，即增强的 before 方法运行在 print 方法之前，after 方法运行在 print 方法之后。

```
before advice
teacher list
teacher list
after advice
```

7.3 Spring 实现 AOP

Spring AOP 的实现既可以基于 XML 方式，也可以基于注解方式，还可以基于 Java 的零配置方式。每一种方式都有其特定的适用场景。下面我们针对每种方式都设计了案例，详细阐述这几种方式的实现过程。

7.3.1 任务三：基于 XML 的 AOP 实现

基于 XML 的 AOP 实现使用的是 org.springframework.aop.framework.ProxyFactoryBean 类，这个类对切点和增强提供了完整的控制功能，并可以生成指定的内容。

ProxyFactoryBean 类中的常用可配置属性如表 7-1 所示。

表 7-1 ProxyFactoryBean 类中的常用属性

属 性 名 称	描 述
target	被代理的目标对象
proxyInterfaces	代理实现的接口，如果有多个接口，则可以使用以下格式赋值： <list> <value ></value> ... </list>
proxyTargetClass	判断是代理类还是接口，设置为 true 时，使用 CGLIB 代理，默认是 true
interceptorNames	织入目标对象的增强名称列表
singleton	返回的代理是否为单例，默认为 true（返回单实例）
optimize	当设置为 true 时，强制使用 CGLIB

【示例 7-3】演示 ProxyFactoryBean 类基于 XML 的使用方法。

步骤 1．在 com.ssm.util 包下新建 Math.java 作为目标类，它有 4 个业务方法——加、减、乘、除，代码如下：

```
package Spring;
/**
 * 被代理的目标类
 */
public class Math {
    private float result;
    public void add(int n1,int n2){
        result=n1+n2;
        System.out.println(result);
    }
    public void subtract(int n1,int n2){
        result=n1-n2;
        System.out.println(result);
    }
    public void mul(int n1,int n2){
```

```
        result=n1*n2;
        System.out.println(result);
    }
    public void dev(int n1,int n2){
        result=n1/n2;
        System.out.println(result);
    }
}
```

步骤 2．在 com.ssm.util 包下新建增强类 Advice.java。代码如下：

```
public class Advice implements MethodBeforeAdvice,AfterReturningAdvice{
    @Override
    public void before(Method method, Object[] args, Object target) throws Throwable
    {
        System.out.println("-----before advice---------");
    }
    @Override
    public void afterReturning(Object returnValue, Method method, Object[] args,
Object target) throws Throwable {
        System.out.println("-----after advice---------");
    }
}
```

Advice 类（增强类）中，定义了两个方法：before 和 after。其中 before 方法是实现 MethodBeforeAdvice 接口的前置方法，而 after 方法是实现 AfterReturningAdvice 接口的后置方法。

步骤 3．在 resources 下新建 aop.xml 文件实现 AOP，代码中有详细的注释。代码如下：

```
    <!--增强-->
    <bean id="advice" class="com.ssm.util.Advice"/>
    <!--被代理的目标类-->
    <bean id="target" class="com.ssm.util.Math"/>
    <!--目标类织入增强后形成的 bean-->
    <bean id="math" class="org.springframework.aop.framework.ProxyFactoryBean"
        p:target-ref="target"
         p:interceptorNames="advice"/>
```

步骤 4．准备一个测试类，获取 aop.xml 文件中的名为 math 的 Bean，并运行其所有的业务方法。代码如下：

```
 @org.junit.Test
    public void test() {
        ApplicationContext applicationContext = new ClassPathXmlApplicationContext
("aop.xml");
        Math math = applicationContext.getBean("math", Math.class);
        int n1 = 100, n2 = 5;
        math.add(n1, n2);
          }
}
```

步骤 5．运行测试方法 Test.test，控制台上输出以下信息：

```
-----before advice---------
105.0
-----after advice---------
```

由此可见，math 对象运行 add 方法时，被成功织入前置增强和后置增强。

如果在示例 7-3 的基础上加入环绕增强，则只需让 Advice 类实现 MethodInterceptor 接口及其 invoke 方法。invoke 方法中的 Object object=invocation.proceed();是执行被代理类的业务方法，被包围在两条输出语句之间。具体实现请参看下面的代码：

```
public class Advice implements
MethodBeforeAdvice,AfterReturningAdvice,MethodInterceptor{
    @Override
    public void before(Method method, Object[] args, Object target) throws Throwable
    {
        System.out.println("-----before advice---------");
    }
    @Override
    public void afterReturning(Object returnValue, Method method, Object[] args,
Object target) throws Throwable {
        System.out.println("-----after advice---------");
    }
    @Override
    public Object invoke(MethodInvocation invocation) throws Throwable {
        System.out.println("-----around advice begin---------");
        Object object=invocation.proceed();
        System.out.println("-----around advice end---------");
        return object;
    }
}
```

修改完毕后，运行测试方法 Test.test，控制台上输出如下信息：

```
-----around advice begin---------
-----before advice---------
105.0
-----after advice---------
-----around advice end---------
```

由上可见，环绕增强将业务方法和前置增强、后置增强都包围了。

7.3.2　任务四：基于注解的 AOP 实现

基于注解的配置，常用以下 5 种注解的增强，前面说的引介增强比较少用。

- @Before：前置。
- @After：后置。
- @Around：环绕。
- @AfterReturnning：运行在 return 之后。
- @AfterThrowing：抛出异常之后。

另外，还要使用@Component 注解定义 Bean，@Aspect 注解定义切面。通常，我们会使用@Before("execution(* Spring.Math.*(..))")结合的方法来定义连接点。

【示例 7-4】在示例 7-3 的基础上稍做修改，实现基于注解的 AOP。

步骤 1. 修改 aop.xml 文件，去掉里面的 Bean 定义和 AOP 配置，替换为自动扫描 com.ssm.util 包，定义 AspectJ 的自动目标类代理。代码如下：

```
<context:component-scan base-package="com.ssm.util">
    </context:component-scan>
<aop:aspectj-autoproxy proxy-target-class="true"></aop:aspectj-autoproxy>
```

步骤 2．修改 Math.java 文件，在类的前面添加注解@Service，定义一个名为 math 的 Bean。

步骤 3．修改 Advice.java 文件，这里的改动比较大，因为主要的切面和切点的注解定义都在这里完成。代码如下：

```
@Component  //定义 Advice 类的 Bean，名为 advice
@Aspect       //定义当前类为一个切面
public class Advice {
    //定义该方法 before 是一个前置方法，切点是 Math 类的所有方法
    @Before("execution(* com.ssm.util.Math.*(..))")
    public void before(){
        System.out.println("-----before advice---------");
}
    //定义该方法 after 是一个后置方法，切点是 Math 类的所有方法
    @After("execution(* com.ssm.util.Math.*(..))")
    public void after(){
        System.out.println("-----after advice---------");    }
}
```

步骤 4．运行测试方法 Test.test，控制台输出跟示例 7-3 一样的结果。

```
-----before advice---------
105.0
-----after advice---------
```

【示例 7-5】在示例 7-4 的基础上实现基于注解的环绕增强。

步骤 1．保持 Math 类不变。

步骤 2．对 Advice 类增加环绕方法和对切点的过滤定义，只作用在以字母 m 开头的业务方法上。为了增强显示效果，对 before 方法和 after 方法的切点也略做修改，分别只作用在以字母 a 和 s 开头的业务方法上。代码如下：

```
    //定义该方法是一个前置方法，切点是 Math 类的 add 方法
    @Before("execution(* com.ssm.util.Math.a*(..))")
    public void before(){
        System.out.println("-----before advice---------");
    }
    //定义该方法是一个后置方法，切点是 Math 类的 subtract 方法
    @After("execution(* com.ssm.util.Math.s*(..))")
    public void after(){
```

```
        System.out.println("-----after advice---------"); }
    @Around("execution(* com.ssm.util.Math.m*(..))")
    private Object around(ProceedingJoinPoint pjp) throws Throwable {
        System.out.println("-----around().invoke-----");
        System.out.println(" 类似于 Before Advice 的操作");
        //调用核心逻辑
        Object retVal = pjp.proceed();
        System.out.println(" 类似于 After Advice 的操作");
        System.out.println("-----End of around()------");
        return retVal;
    }
```

步骤 3．保持 aop.xml 文件不变。

步骤 4．对测试方法 Test.test 稍做修改，为了显示不同增强方法的织入，增加对 math 对象的减法、乘法、除法的调用。

步骤 5．运行测试方法 Test.test，控制台上输出如下信息：

```
-----before advice---------
105.0
95.0
-----after advice---------
-----around().invoke-----
 类似于 Before Advice 的操作
500.0
 类似于 After Advice 的操作
-----End of around()------
20.0
```

由上可见，前置增强方法 before 只运行在 add 方法之前，后置增强方法 after 运行在 subtract 方法之后，而环绕增强方法 around 运行在 mul 方法的周围，dev 方法中没有任何增强方法运行。

7.4 AspectJ 函数和其他 AOP 的实现

Spring 自 2.0 版本开始支持 AspectJ 框架，可以使用 AspectJ 专门的切点表达式描述切面。Spring 所支持的 AspectJ 表达式函数分为 4 类：

- 方法切点函数：通过描述目标类方法信息定义连接点。
- 方法参数切点函数：通过描述目标类方法入参信息定义连接点。
- 目标类切点函数：通过描述目标类类型信息定义连接点。
- 代理类切点函数：通过描述代理类类型信息定义连接点。

常见的 AspectJ 表达式函数如下：

execution()：满足匹配模式字符串的所有目标类方法的连接点。

@annotation()：任何标注了指定注解的目标类方法的连接点。

args()：目标类方法运行时指定了参数的类型的连接点。

@args()：目标类方法参数中是否有指定注解的连接点。

within()：匹配指定的包的所有连接点。

@within()：匹配目标对象拥有指定注解的类的所有连接点。

target()：匹配指定目标类的所有连接点。

@target()：匹配当前目标对象类型的执行连接点，其中目标对象持有指定的注解。

this()：匹配当前 AOP 代理对象类型的所有执行连接点。

最常用的是 execution(<修饰符模式>?<返回类型模式><方法名模式>(<参数模式>)<异常模式>?)切点函数，可以满足大多数需求。在前面的示例中都使用了 execution 函数。

下面我们讲解 AspectJ 函数注解@annotation 和@target 的使用方法。

7.4.1 任务五：@annotation 自定义注解的使用

下面我们通过实现自定义注解类来了解@annotation 注解的使用方式。

【示例 7-6】实现自定义注解类。

步骤 1．实现自定义注解类 MyAnno。代码如下：

```
import java.lang.annotation.*;
@Target({ElementType.METHOD})
@Retention(RetentionPolicy.RUNTIME)
@Documented
public  @interface MyAnno
{}
```

@Target、@Retention、@Documented 都是 Java 5.0 定义的元注解。

@Target 说明了 Annotation 所修饰的对象范围。Annotation 应用的范围很广，包括 packages、types（类、接口、枚举、Annotation 类型）、类型成员（方法、构造方法、成员变量、枚举值）、方法参数和本地变量（如循环变量、catch 参数）。在 Annotation 类型的声明中使用@Target 可明确其修饰的目标。在上面的代码中，MyAnno 注解被声明用于方法上。

@Retention 定义了 Annotation 被保留的时长。某些 Annotation 仅出现在源代码中，而被编译器丢弃；而另一些却被编译在 class 文件中；编译在 class 文件中的 Annotation 可能会被虚拟机忽略，而另一些在 class 被装载时将被读取（请注意并不影响 class 的执行，因为 Annotation 与 class 在使用上是被分离的）。使用这个元注解可以限制 Annotation 的生命周期。在上面的代码中，MyAnno 注解的生命周期被定义为 runtime，即运行时有效。

@Documented 注解表明这个注解会被类似 javadoc 的工具记录。默认情况下，javadoc 记录是不包括注解的。但如果声明注解时标识了@Documented，则它会被 javadoc 之类的工具记录，所以注解类型信息也会被包括在生成的文档中。Documented 是一个标记注解，没有成员。

步骤 2．修改后置增强方法 after，使之运行在标注了 MyAnno 注解的方法之后。

After 方法实现了输出 MyAnno advice 和连接点的方法名。代码如下：

```
@After("@annotation(com.ssm.util.MyAnno)")
   public void after(JoinPoint joinPoint)
   {
       System.out.println("----MyAnno  advice------");
       System.out.println(joinPoint.getSignature().getName());
   }
```

步骤 3．修改 Math 方法的 mul 方法，为其添加@MyAnno 注解，使其调用后置增强方法。代码如下：

```
@MyAnno
   public void mul(int n1,int n2){
       result=n1*n2;
       System.out.println(result);
   }
```

步骤 4．其他无须修改，直接运行测试方法 Test.test，控制台上输出如下信息：

```
-----before advice---------
105.0
95.0
-----around().invoke-----
 类似于 Before Advice 的操作
500.0
 类似于 After Advice 的操作
-----End of around()------
----MyAnno  advice------
mul
20.0
```

由上面的代码看出，mul 方法除了被 around 方法围绕，还在其后运行了 after 方法，输出了----MyAnno advice------和方法名称 mul。这说明@MyAnno 注解起到了作用。

7.4.2 任务六：@target 注解的使用

在前文中提到@target 注解会匹配当前目标对象类型的执行方法，其中目标类上标有指定的注解。

【示例 7-7】演示@target 注解的使用方法。

步骤 1．将 MyAnno 类的@Target({ElementType.METHOD})改为@Target({ElementType.TYPE})，表明该注解可以定义在类上。代码如下：

```
import java.lang.annotation.*;
@Target({ElementType.TYPE})
@Retention(RetentionPolicy.RUNTIME)
@Documented
public  @interface MyAnno
{}
```

步骤 2．修改 Advice 类的 after 方法，代码如下：

```
@After("execution(* com.ssm.util.Math.m*(..))&&@target(com.ssm.util.MyAnno)")
   public void after(JoinPoint joinPoint)
   {
       System.out.println("----MyAnno  advice------");
       System.out.println(joinPoint.getSignature().getName());
}
```

以前使用@After("@annotation(com.ssm.util.MyAnno)")注解来定义切点，现在改为

@After("execution(* com.ssm.util.Math.m*(..))&&@target(com.ssm.util.MyAnno)")，切点是 Math 类上以字母 m 开头的方法。如果 Math 类上没有 MyAnno 注解，则 execution 函数也会无效。

步骤 3．修改 Math 类，把注解@MyAnno 从 mul 方法移到类的声明上。代码如下：

```
@Service
@MyAnno
public class Math {...}
```

步骤 4．运行 Test.test，控制台上输出效果跟示例 7-6 完全一样。如果注释掉步骤 3 中的 @MyAnno，则 After 方法不被执行。代码如下：

```
-----before advice---------
105.0
95.0
-----around().invoke-----
 类似于 Before Advice 的操作
500.0
 类似于 After Advice 的操作
-----End of around()------
20.0
```

7.4.3 任务七：自动创建代理

在 7.3.1 节中，Spring AOP 通过类 ProxyFactoryBean 创建代理对象，它有个缺陷，就是只能代理一个目标对象 Bean。当代理目标类过多时，配置文件臃肿，并且管理和维护特别麻烦。因此，Spring 提供了能够让容器自动创建代理的类 BeanNameAutoProxyCreator、Default AdvisorAutoProxyCreator。下面演示二者是如何实现自动创建代理的。

【示例 7-8】使用 BeanNameAutoProxyCreator 实现自动创建代理。

步骤 1．准备 Math 和 Car 类。

步骤 2．修改 Advice 类，定义并实现了 MethodBeforeAdvice、AfterReturningAdvice 接口。代码如下：

```
public class Advice implements MethodBeforeAdvice,AfterReturningAdvice{
    @Override
    public void before(Method method, Object[] args, Object target) throws Throwable
    {
        System.out.println("-----before advice---------");
    }
    @Override
    public void afterReturning(Object returnValue, Method method, Object[] args,
Object target) throws Throwable {
        System.out.println("-----after advice---------");
    }
}
```

步骤 3．修改 aop.xml 文件，改为基于 XML 的自动创建代理的配置方式，里面两个 Bean 都织入了增强。代码如下：

```
    <bean id="advice" class="com.ssm.util.Advice"/>
    <bean id="math" class="com.ssm.util.Math"/>
```

```
    <bean id="car" class="com.ssm.util.Car">
        <property name="maxSpeed" value="500"/>
        <property name="type" value="BMW"/>
    </bean>
    <bean    class="org.springframework.aop.framework.autoproxy.BeanNameAutoProxy
Creator"
        p:beanNames="math,car"
        p:interceptorNames="advice"
        p:optimize="true"/>
```

步骤 4．修改 Test 类文件，调用 math 对象和 car 对象的业务方法。代码如下：

```
  @org.junit.Test
    public void test() {
        ApplicationContext applicationContext = new ClassPathXmlApplicationContext
("aop.xml");
        Math math = applicationContext.getBean("math", Math.class);
        Car car=applicationContext.getBean("car",Car.class);
        int n1 = 100, n2 = 5;
        math.add(n1, n2);
        car.introduce();
    }
```

步骤 5．运行 Test.test，控制台输出如下信息：

```
  -----before advice---------
  105.0
  -----after advice---------
  -----before advice---------
  type is: BMW ,maxSpeed is: 500
  -----after advice---------
```

由上可见，无论是 math 对象，还是 car 对象，都调用了前置增强方法和后置增强方法，成功实现了自动创建代理。

使用 DefaultAdvisorAutoProxyCreator 可以将增强和切点织入，配置代码如下：

```
  <bean id="advice" class="com.ssm.util.Advice"/>
  <bean id="math" class="com.ssm.util.Math"/>
  <bean id="car" class="com.ssm.util.Car">
     <property name="maxSpeed" value="500"/>
     <property name="type" value="BMW"/>
  </bean>
  <bean  id="regexpAdvisor2"  class="org.springframework.aop.support.RegexpMethod
PointcutAdvisor"
       p:pattern="com.ssm.util.*"
       p:advice-ref="advice"/>
  <bean class="org.springframework.aop.framework.autoproxy.DefaultAdvisorAutoProxy
Creator"/>
```

7.4.4　任务八：基于 Schema 的 AOP 实现

如果项目不能使用 Java 5.0，就无法使用 7.3.2 节中的基于注解的切面。 但是 Spring 提供了基于 Schema 配置的方法，它完全可以替代基于注解声明切面的方式。

基于注解的切面，就是将切点、增强类型的信息使用注解描述。我们只要将这些信息转移到 Schema 的 XML 配置文件中就可以了。使用 Schema 定义切面，切点、增强类型的注解信息会从切面类中剥离出来。

在 Spring 配置文件中，所有 AOP 相关定义必须放在<aop:config>标签下，该标签下可以有<aop:pointcut>、<aop:advisor>、<aop:aspect>、<aop:before>等标签，配置顺序可按一定的规则变化。

<aop:pointcut>：用来定义切点，该切点可以重用。

<aop:advisor>：用来定义只有一个增强和一个切点的切面。

<aop:aspect>：用来定义切面，该切面可以包含多个切点和增强，而且标签内部的增强和切点定义是无序的。

【示例 7-9】使用 Schema 实现 AOP 的前置、后置增强。

步骤 1．Math 类保持不变，仍然使用示例 7-3 中的 Math 类。

步骤 2．修改 Advice 类，其代码非常简洁，如下所示。

```
public class Advice {
   public void before(){
       System.out.println("-----before advice---------");
   }
   public void after(){
       System.out.println("-----after advice---------");
          }
}
```

Advice 类中，定义了两个方法 before 和 after。其中 before 方法将会在配置时被定义成前置增强方法，而 after 方法将会被定义成后置增强方法。

步骤 3．在 resources 下新建 aop.xml 文件实现 AOP，代码中有详细的注释，内容如下：

```
<!-- 被代理对象 math -->
   <bean id="math" class="com.ssm.util.Math"></bean>
   <!-- 增强 advice -->
   <bean id="advice" class="com.ssm.util.Advice"></bean>
   <!-- aop 配置,设置成目标代理类，如果 proxy-target-class 设置为 false，则需要接口 -->
   <aop:config proxy-target-class="true">
       <!--切面引用 advice 增强 -->
       <aop:aspect ref="advice">
           <!-- 切点 pointcut1,Math 的所有业务方法都是连接点 -->
           <aop:pointcut  expression="execution(*  com.ssm.util.Math.*(..))"  id=
"pointcut1"/>
           <!--连接增强方法与切点 -->
           <!--advice 的 before 方法作为切点 pointcut1 的前置方法 -->
           <aop:before method="before" pointcut-ref="pointcut1"/>
           <!--advice 的 after 方法作为切点 pointcut1 的后置方法 -->
           <aop:after method="after" pointcut-ref="pointcut1"/>
       </aop:aspect>
  </aop:config>
```

步骤 4．准备一个测试类，获取 aop.xml 文件中的 math Bean，并运行其所有的业务方法。

代码如下：

```
public class SpringTest {
    @Test
    public void maint(){
        ApplicationContext applicationContext=new ClassPathXmlApplicationContext
("aop.xml");
        Math math = applicationContext.getBean("math", Math.class);
        int n1 = 100, n2 = 5;
        math.add(n1, n2);
        math.subtract(n1, n2);
        math.mul(n1, n2);
        math.dev(n1, n2);    }
```

步骤 5．运行 Test.test，控制台上输出以下内容：

```
-----before advice---------
105.0
-----after advice---------
-----before advice---------
95.0
-----after advice---------
-----before advice---------
500.0
-----after advice---------
-----before advice---------
20.0
-----after advice---------
```

可以看出，math 对象的每个业务方法前后都分别调用了 Advice 类的前置方法和后置方法。

基于 Schema 定义 AOP 很灵活，也可以将切点跟增强分离，效果不变。代码修改如下：

```
<!-- aop 配置 -->
    <aop:config proxy-target-class="true">
        <!-- 切点 -->
        <aop:pointcut expression="execution(* com.ssm.util.Math.*(..))" id=
"pointcut1"/>
        <!--切面 -->
        <aop:aspect ref="advice">
            <!--连接增强方法与切点 -->
            <aop:before method="before" pointcut-ref="pointcut1"/>
            <aop:after method="after" pointcut-ref="pointcut1"/>
        </aop:aspect>
    </aop:config>
```

如果将后置增强定义

```
<aop:after method="after" pointcut-ref="pointcut1"/>
```

改成

```
<aop:after-returning method="after" pointcut-ref="pointcut1"/>
```

运行效果也不会变。那么 after 和 after-returning 到底有什么区别呢？看下面的代码段和注释就会明白了。

```
try{
    try{
        //前置@Before
        method.invoke(..);
    }finally{
        //后置@After
    }
    //带 return 参数@AfterReturning
}catch(){
    //异常@AfterThrowing
}
```

接下来，在示例 7-9 的基础上，再示范一个基于 Schema 的实现环绕增强的案例。

【示例 7-10】使用 Schema 实现环绕增强。

步骤 1．Math 类保持不变。

步骤 2．为 Advice 类添加环绕方法的实现代码。代码如下：

```
public class Advice {
    public void before(){
        System.out.println("-----before advice---------");
    }
    public void after(){
        System.out.println("-----after advice---------");
    }
    private Object around(ProceedingJoinPoint pjp) throws Throwable {
        System.out.println("-----around().invoke-----");
        System.out.println(" 类似于 Before Advice 的操作");
        //调用核心逻辑
        Object retVal = pjp.proceed();
        System.out.println(" 类似于 After Advice 的操作");
        System.out.println("-----End of around()------");
        return retVal;
    }
}
```

步骤 3．修改 aop.xml 文件，定义环绕增强的切面。代码如下：

```
    <!-- aop 配置 -->
    <aop:config proxy-target-class="true">
        <!-- 切点 -->
        <aop:pointcut   expression="execution(*   com.ssm.util.Math.*(..))"   id=
"pointcut1"/>
        <!--切面 -->
        <aop:aspect ref="advice">
            <!--连接增强 around 方法与切点 -->
            <aop:around method="around" pointcut-ref="pointcut1"/>
        </aop:aspect>
</aop:config>
```

步骤 4．Test 类保持不变，运行 Test.test，控制台输出如下信息：

```
-----around().invoke-----
 类似于 Before Advice 的操作
105.0
 类似于 After Advice 的操作
-----End of around()------
-----around().invoke-----
 类似于 Before Advice 的操作
95.0
 类似于 After Advice 的操作
-----End of around()------
-----around().invoke-----
 类似于 Before Advice 的操作
500.0
 类似于 After Advice 的操作
-----End of around()------
-----around().invoke-----
 类似于 Before Advice 的操作
20.0
 类似于 After Advice 的操作
-----End of around()------
```

可以看出，Math 类的业务方法被 Advice 类的 around 方法包围了。

如果我们不希望将 Math 类的所有方法都当作连接点，可以在 aop:pointcut 的 expression 中使用正则表达式实现过滤。例如，只对以字母 a 开头的方法实现增强，可以这样写：

```
    <aop:pointcut    expression="execution(*    com.ssm.util.Math.a*(..))"    id=
"pointcut1"/>
```

运行效果如下：

```
-----around().invoke-----
 类似于 Before Advice 的操作
105.0
 类似于 After Advice 的操作
-----End of around()------
95.0
500.0
20.0
```

由上可见，只有 add 方法被当作连接点实现了环绕增强。

7.4.5 任务九：零配置实现 AOP

所谓零配置实现 Spring AOP，即在基于注解的基础上，去掉 aop.xml 文件。那么，aop.xml 文件中定义的扫描包的定义和自动代理在哪儿完成呢？答案是将其化为注解置于运行类上。

【示例 7-11】实现零配置 AOP。

步骤 1．为 Math 类添加注解@Service。

步骤 2．修改 Advice 类，还原为 7.3.2 节中基于注解的形式。

步骤 3．修改 Test 类，零配置实现的要点就在这个类上。代码如下：

```
//用于表示当前类为容器的配置类，类似<beans/>
```

```
    @Configuration
    //扫描的范围，相当于 XML 配置的结点<context:component-scan/>
  @ComponentScan(basePackages="com.ssm.util")
    //自动代理，相当于<aop:aspectj-autoproxy proxy-target-class="true"></aop:aspectj-
autoproxy>
  @EnableAspectJAutoProxy(proxyTargetClass=true) public class Test {
      @org.junit.Test
      public void test() {
          ApplicationContext   applicationContext=new   AnnotationConfigApplication
Context(Test.class);
          Math math = applicationContext.getBean("math", Math.class);
          int n1 = 100, n2 = 5;
          math.add(n1, n2);
          math.subtract(n1, n2);
          math.mul(n1, n2);
          math.dev(n1, n2);
      }
  }
```

我们可以看到，在 Test 类上多了三个注解，并对该注解给出非常详细的解释说明。在 test 方法里面，不再使用 ClassPathXmlApplicationContext 类来读取 aop.xml 文件，而是使用 AnnotationConfigApplicationContext 类读取基于注解的类。这样，就实现了零配置 AOP。

步骤 4．运行 Test.test，控制台输出如下信息：

```
-----before advice---------
105.0
-----after advice---------
-----before advice---------
95.0
-----after advice---------
-----before advice---------
500.0
-----after advice---------
-----before advice---------
20.0
-----after advice---------
```

7.5 Spring 声明式事务

Spring 事务管理是 Spring AOP 技术的精彩应用，它作为一个切面织入目标业务方法，使得业务代码从事务代码中脱离出来。Spring 为事务管理提供了一致的编程模板，不管是 Spring JDBC、Hibernate、JPA 还是 MyBatis，用户都可以使用统一的编程模型进行事务管理。

Spring 支持声明式事务管理和编程式事务管理两种方式。声明式事务管理在 AOP 上实现，事务的本质是增强 Advice，拦截目标类的业务方法，然后在目标方法开始之前创建或者织入一个事务，在执行完目标方法之后根据执行情况提交或者回滚事务。声明式事务的配置非常方便，只需要提供切点信息和控制事务行为的属性信息。声明式事务跟 AOP 一样，不需要通过编程的方式管理事务，不需要在业务逻辑代码中掺杂事务管理的代码，只需在配置文件中进行相关的事务规则声明（或通过基于@Transactional 注解的方式），便可以将事务规则应用到业务逻辑中。声明式事务管理使项目高度模块化，减少了代码冗余，提高了代码的可读性。

显然，声明式事务管理要优于编程式事务管理，因为它恰好是 Spring 倡导的非侵入式的开发方式。声明式事务管理使业务代码保持完整，一个普通的 POJO，只要基于 XML 配置或者加上注解就可以获得完全的事务支持。和编程式事务相比，声明式事务唯一不足的地方是，它只能作用到方法级别，无法像编程式事务那样作用到代码块级别。但是即便这样，也存在很多变通的方法，如可以将需要进行事务管理的代码块独立为方法等。

7.5.1 Spring 声明式事务特性

Spring 所有的事务管理策略类都继承自 org.springframework.transaction.PlatformTransaction Manager 接口。Spring 事务策略是通过 PlatformTransactionManager 接口实现的，它是整个 Spring 事务的核心。该接口对事务策略高度抽象，不依赖于任何具体的事务策略。对于底层具体的事务策略，它有相应的实现类。不同的事务策略的切换，通常由 Spring 容器负责管理。其业务功能不必与具体的事务 API 耦合，也无须与特定的事务实现类耦合。

其中，TransactionDefinition 接口定义了以下事务特性：

1．事务的传播行为

Spring 提供了 7 种类型的事务传播行为，可以通过事务的 propagation 属性来进行配置：

- PROPAGATION_REQUIRED：支持当前事务，如果当前没有事务，就新建一个事务。这是最常见的选择。
- PROPAGATION_SUPPORTS：支持当前事务，如果当前没有事务，就以非事务方式执行。
- PROPAGATION_MANDATORY：支持当前事务，如果当前没有事务，就抛出异常。
- PROPAGATION_REQUIRES_NEW：新建事务，如果当前存在事务，就把当前事务挂起。
- PROPAGATION_NOT_SUPPORTED：以非事务方式执行操作，如果当前存在事务，就把当前事务挂起。
- PROPAGATION_NEVER：以非事务方式执行操作，如果当前存在事务，则抛出异常。

2．事务并发会产生的问题

- A 脏读：如事务 A 读到事务 B 未提交的数据。
- B 不可重复读：如多次读取同一个事务得到的结果不同。
- C 幻读：如事务 A 读到事务 B 已提交的数据。

3．事务的隔离级别

事务的隔离级别是指若干个并发的事务之间的隔离程度，如表 7-2 所示。

表 7-2 事务隔离级别

隔 离 级 别	解 释	允许并发的问题
ISOLATION_DEFAULTD	使用数据库默认的事务隔离级别	
ISOLATION_READ_UNCOMMITTED	最低的隔离级别，它允许另外一个事务看到这个事务未提交的数据	A、B、C

续表

隔 离 级 别	解 释	允许并发的问题
ISOLATION_READ_COMMITTED	保证一个事务修改的数据提交后才能被另一个事务读取。另一个事务不能读取该事务未提交的数据	B、C
ISOLATION_REPEATABLE_READ	保证一个事务不能读取另一个事务未提交的数据	C
ISOLATION_SERIALIZABLE	花费代价最高，但是最可靠的事务隔离级别	none

注：A、B、C 为上文 2. 中出现的问题。

7.5.2 事务的配置方式

声明式事务管理有两种常用的方式，一种是基于 tx 和 aop 命名空间的 XML 文件配置方式，另一种是基于@Transactional 注解的方式。显然基于注解的方式更简单易用。

1．基于 tx 和 aop 命名空间的 XML 文件配置方式

Spring 配置文件中，事务配置由三个部分组成，分别是 DataSource、TransactionManager 和代理机制。DataSource、TransactionManager 会根据数据访问方式有所变化。例如，在我们的项目中使用 MyBatis 进行数据访问时，DataSource 实际上是 org.Mybatis.Spring.SqlSessionFactoryBean，TransactionManager 的实现则为 org.Springframework.jdbc.datasource.DataSourceTransactionManager。

Spring 在 Schema 的基础上，添加了 tx 命名空间，明确以结构化的方式定义事务属性，配合 aop 命名空间所提供的切面定义，使得业务类方法事务的配置大大简化。配置代码如下：

```
    <!-- 指定事务管理器 -->
    <bean id="txManager" class="org.springframework.jdbc.datasource. DataSourceTransaction
Manager">
    <property name="dataSource" ref="dataSource"></property>
    </bean>
    <!-- 设置事务增强 -->
    <tx:advice id="txAdvice" transaction-manager="txManager">
    <tx:attributes>
        <tx:method name="get*" read-only="true"/>
        <tx:method name="add*,update*" rollback-for="Exception"/>
        </tx:attributes>
        </tx:advice>
    <!-- 作用 Schema 的方式配置事务,这里是把事务设置到 service 层-->
        <aop:config>
        <aop:pointcut  id="servicePointcut"  expression="execution(*  com.ssm.service.
*(..))" />
        <aop:advisoradvice-ref="txAdvice"pointcut-ref="servicePointcut"/>
        </aop:config>
```

上面的代码创建了事务管理器 Bean，名称是 txManager，由 DataSourceTransactionManager 类创建。随后，将 txManager 定义为增强 txAdvice，通过 attributes 属性定义以 get 开头的方法都是只读方法，以 add、update 开头的方法在出现异常时回滚。接下来，将增强 txAdvice 织入业务，切点是 com.ssm.service 包下所有类的所有方法。

【示例 7-12】演示基于 XML 的声明式事务配置方式，其间用到了附录 A 中数据库的教师表 teacher：

步骤 1．修改 TeacherServiceImpl.java 文件，将其当作将要测试事务管理的业务目标类，代码如下：

```
public class TeacherServiceImpl{
    @Autowired
    private DataSource dataSource;
    public DataSource getDataSource() {
        return dataSource;
    }
    public void setDataSource(DataSource dataSource) {
        this.dataSource = dataSource;
    }
    public void delete(int id){
        try {
            Connection connection=dataSource.getConnection();
            String sql="delete from teacher where id=?";
            PreparedStatement ps=connection.prepareStatement(sql);
            ps.setInt(1,id);
            ps.execute();
            if(id==1)throw new RuntimeException("rollback test") ;
            else
                connection.commit();
        } catch (SQLException e) {
            e.printStackTrace();
        }
    }
}
```

上面的代码中，dataSource 属性是被注入的 Bean，delete 方法通过 id 删除对应的 teacher 记录，里面用到了预编译语句。这里值得注意的是，为了表现声明式事务的织入，这里假设抛出了一个异常。如果 id=1，则抛出 RuntimeException 异常，回滚删除操作，否则提交删除操作。

步骤 2．修改 aop.xml 文件，代码里有详细的解释，内容如下：

```
<!--定义 DriverManagerDataSource 类创建的 Bean-dataSource-->
<bean id="dataSource"
    class="org.springframework.jdbc.datasource.DriverManagerDataSource">
    <property name="driverClassName" value="com.mysql.jdbc.Driver"/>
    <property name="url" value="jdbc:mysql://localhost:3306/test"/>
    <property name="username" value="root"/>
    <property name="password" value="123"/>
</bean>
<!-- 初始化 TransactionManager 类创建的 BeantransactionManager，引用上面的
dataSource- -->
<bean id="transactionManager"
    class="org.springframework.jdbc.datasource.DataSourceTransactionManager">
    <property name="dataSource" ref="dataSource" />
```

```
        </bean>
        <!-- 定义 teacherServiceImpl Bean -->
        <bean id="teacherServiceImpl"class="com.ssm.service.impl.TeacherServiceImpl">
            <property name="dataSource"  ref="dataSource" />
        </bean>
        <!--定义增强，使用前面的事务管理 Bean-transactionManager-->
        <tx:advice id="txAdvice"  transaction-manager="transactionManager">
            <tx:attributes>
                <!--定义 delete 方法，遇到 Exception 回滚-->
                <tx:method name="delete" rollback-for="Exception.class"/>
            </tx:attributes>
        </tx:advice>
        <!--定义切面，切点是 com.ssm.service.impl 包下所有类的所有方法，增强是前面定义的
txAdvice-->
        <aop:config>
            <aop:pointcut id="createOperation"
                    expression="execution(* com.ssm.service.impl.*.*(..))"/>
            <aop:advisor advice-ref="txAdvice" pointcut-ref="createOperation"/>
        </aop:config>
```

步骤 3．修改 Test.test 方法，读取 aop.xml 文件中定义的 Bean、teacherServiceImpl，删除 id=1 的记录。

```
    public class Test {
        @org.junit.Test
        public void test() {
            ApplicationContext context =
                    new ClassPathXmlApplicationContext("aop.xml");
            TeacherServiceImpl teacherServiceImpl =
                    (TeacherServiceImpl)context.getBean("teacherServiceImpl");
            teacherServiceImpl.delete(1);
            }
 }
```

步骤 4．在数据库表 teacher 里添加 id=1 和 id=4 的记录，效果如图 7-3 所示。

Id	name	password	sex	birthday	course_id	professional	salary
1	(Null)	(Null)	(Null)	(Null)	(Null)	(Null)	(Null)
4	(Null)	(Null)	(Null)	(Null)	(Null)	(Null)	(Null)
201	李青山	0000	1	1965-01-01	1	教授	3000
202	唐嫣然	0000	1	1968-01-01	2	教授	8000
203	萧玄茂	0000	1	1978-01-01	3	高级教师	9008

图 7-3　在 teacher 表里添加 id=1、id=4 的记录

提示：InnoDB 引擎的 MySQL 数据库的 autocommit（自动提交）属性的值默认为 1，会导致回滚无效。所以，要将数据库的 autocommit 属性值改为 0。如何改呢？

首先停止 MySQL 服务器，在 CMD 下使用命令 sc delete MySQL 删除服务，然后修改 my.ini 文件，添加 autocommit=0，即修改为手动提交，然后在 CMD 下运行 mysql -install、net start mysql，重新安装并启动 MySQL 数据库服务器。my.ini 文件请参考教学资源。

改为手动提交后，在 Navicat 后台使用如下 SQL 语句插入两条新记录作为后面的测试数据：

```
insert into teacher(id)  values(1);
insert into teacher(id) values(4);
commit;
```

步骤 5. 运行 Test.test，观察删除 id=1 记录的结果，控制台输出如下信息：

```
java.lang.RuntimeException: rollback test
...
Process finished with exit code -1
```

这说明抛出了 RuntimeException 异常，回滚发生了。检查数据库，会发现 id=1 的记录仍然存在，说明事务管理的织入对业务 Bean 起到了作用。

步骤 6. 修改 Test.test 方法，将 teacherServiceImpl.delete(1)改为 teacherServiceImpl.delete(4)，再运行，控制台输出如下信息：

```
Process finished with exit code 0
```

说明程序正常运行，检查数据库，id=4 的记录已被删除，id=1 的记录仍然存在，如图 7-4 所示。以上的操作步骤证明了事务配置的有效性。

Id	name	password	sex	birthday	course_id	professional	salary
1	(Null)	(Null)	(Null)	(Null)	(Null)	(Null)	(Null)
201	李青山	0000	1	1965-01-01	1	教授	3000
202	唐嫣然	0000	1	1968-01-01	2	教授	8000
203	萧玄茂	0000	1	1978-01-01	3	高级教师	9008

图 7-4　执行删除操作后的 teacher 表

2. 基于@Transactional 注解的方式

除了基于 XML 的事务配置，Spring 还提供了基于注解的事务配置，即通过@Transactional 对需要进行事务增强的 Bean 接口实现类或方法进行标注。在容器中配置基于注解的事务增强驱动，即可启用基于注解的声明式事务。

将基于 XML 文件配置方式的 aop 配置替换成如下代码：

```
 <!-- 打开 tx 事务注解管理功能 -->
<tx:annotation-driven transaction-manager="transactionManager"/>
```

然后在业务 Bean 中使用@Transactional 注解，如图 7-5 所示。

```
applicationContext.x    UserDAOImpl.java    UserService.java
package dao;
import java.sql.ResultSet;


@Transactional
@Repository
public class UserDAOImpl  {
    private JdbcTemplate jdbcTemplate;

```

图 7-5　事务注解

【示例 7-13】基于注解的事务配置的使用。

在示例 7-12 的基础上进行如下修改。

步骤 1．修改 aop.xml 文件，删掉 aop 的标签，换成注解驱动事务管理。代码如下：

```
<!--定义 DriverManagerDataSource 类创建的 Bean-dataSource-->
    <bean id="dataSource"
        class="org.springframework.jdbc.datasource.DriverManagerDataSource">
        <property name="driverClassName" value="com.mysql.jdbc.Driver"/>
        <property name="url" value="jdbc:mysql://localhost:3306/test"/>
        <property name="username" value="root"/>
        <property name="password" value="123"/>
    </bean>
    <!-- 初始化 TransactionManager 类创建的 BeantransactionManager, 引用上面的
dataSource- -->
    <bean id="transactionManager"
        class="org.springframework.jdbc.datasource.DataSourceTransactionManager">
        <property name="dataSource"  ref="dataSource" />
    </bean>
    <!-- 定义 teacherServiceImpl Bean -->
    <bean id="teacherServiceImpl"
        class="com.ssm.service.impl.TeacherServiceImpl">
        <property name="dataSource"  ref="dataSource" />
    </bean>
    <!--定义注解驱动的事务管理，使用 Bean-transactionManager-->
   <tx:annotation-driven transaction-manager="transactionManager"/>
```

步骤 2．修改 TeacherServiceImpl.java 文件，在类上添加@Transactional 注解。

步骤 3．Test.test 不变，以跟示例 7-12 一样的步骤执行，首先在 teacher 表里添加一条 id=4 的记录并手动提交。然后删除 id=1 的记录，控制台输出回滚的异常信息，再删除 id=4 的记录，控制台输出正常执行的信息。最后，检查数据库表 teacher，发现 id=1 的记录还在，id=4 的已然被删除。同样，这也证明了事务配置的有效性。

7.5.3 项目中使用 Spring AOP 实现数据库的事务管理

由 SSM 框架集成的项目，其数据库的事务管理在 spring-mybatis.xml 文件中配置。代码如下：

```
    <!-- (事务管理)transactionManager,使用 DataSourceTransactionManager 类创建的 Bean,
注入 Bean-dataSource -->
    <bean id="transactionManager"
        class="org.springframework.jdbc.datasource.DataSourceTransactionManager">
        <property name="dataSource" ref="dataSource" />
    </bean>
    <!--定义增强，使用前面的事务管理 Bean-transactionManager-->
    <tx:advice id="txAdvice" transaction-manager="transactionManager">
        <tx:attributes>
            <!--定义 get 字符开头的方法，只读事务-->
            <tx:method name="get*" read-only="true"/>
```

```
            <!--定义 insert 字符开头的方法，遇到 Exception 回滚-->
            <tx:method name="insert*" rollback-for="Exception"/>
            <!--定义 update*,delete*方法-->
            <tx:method name="update*,delete*"/>
        </tx:attributes>    </tx:advice>
    <!--使用 Schema 方式定义切面，切点是 com.ssm.service.impl 包下所有类的所有方法，增强是前面定义的 txAdvice-->
    <aop:config>
        <aop:pointcut  id="serviceOperation"                                expression=
"execution(* com.ssm.service.impl.*.*(..))" />
        <aop:advisor advice-ref="txAdvice" pointcut-ref="serviceOperation" />
    </aop:config>
```

7.6　实现三大框架总集成的配置文件

至此为止，我们就完成了集成三大框架的总框架，在这里贴出最后的配置文件。大家可以参照此文件检查自己课程设计的配置。源码在教学资源第 7 章 7-14 里。

1．项目根目录下的 build.gradle 文件

build.gradle 文件内容如下：

```
group 'person.xjl'
version '1.0-SNAPSHOT'
apply plugin: 'java'
apply plugin: 'war'
sourceCompatibility = 1.8
configurations {
    MybatisGenerator
}
repositories {
    maven {
        name "aliyunmaven"
        url "http://maven.aliyun.com/nexus/content/groups/public/"
    }
    mavenCentral()
}
dependencies {
 testCompile group: 'junit', name: 'junit', version: '4.11'
    // https://mvnrepository.com/artifact/org.springframework/Spring-web
compile group: 'org.springframework', name: 'spring-web', version: '4.3.18.RELEASE'
// https://mvnrepository.com/artifact/org.springframework/spring-webmvc
compile   group:   'org.springframework',   name:   'spring-webmvc',   version:
'4.3.18.RELEASE'
    // https://mvnrepository.com/artifact/javax.servlet/javax.servlet-api
 compile group: 'javax.servlet', name: 'javax.servlet-api', version: '3.1.0'
 // https://mvnrepository.com/artifact/javax.servlet/jstl
 compile group: 'javax.servlet', name: 'jstl', version: '1.2'
```

```
    // https://mvnrepository.com/artifact/org.hibernate/hibernate-validator
    compile group: 'org.hibernate', name: 'hibernate-validator', version:
'5.4.1.Final'
    //Mybatis
    compile "org.Mybatis:Mybatis:3.4.1"
    //Mybatis Spring 插件
    compile "org.Mybatis:Mybatis-Spring:1.3.1"
    // https://mvnrepository.com/artifact/mysql/mysql-connector-java
    compile group: 'mysql', name: 'mysql-connector-java', version: '5.1.18'
    // https://mvnrepository.com/artifact/org.springframework/Spring-jdbc
    compile group: 'org.springframework', name: 'spring-jdbc', version:
'4.3.18.RELEASE'
    // https://mvnrepository.com/artifact/org.springframework/spring-tx
    compile group: 'org.springframework', name: 'spring-tx', version: '4.3.18.RELEASE'
    // https://mvnrepository.com/artifact/log4j/log4j
      compile group: 'log4j', name: 'log4j', version: '1.2.17'
    //公共资源包
      compile "commons-logging:commons-logging:1.2"
      compile "commons-lang:commons-lang:2.6"
      compile "org.apache.commons:commons-collections4:4.0"
      compile "commons-beanutils:commons-beanutils:1.8.3"
      compile "commons-dbcp:commons-dbcp:1.4"
      compile "commons-pool:commons-pool:1.6"
      // https://mvnrepository.com/artifact/com.mchange/c3p0
      compile group: 'com.mchange', name: 'c3p0', version: '0.9.5.2'
      // https://mvnrepository.com/artifact/com.alibaba/druid
      compile group: 'com.alibaba', name: 'druid', version: '1.1.10'
      //mvnrepository.com/artifact/com.github.pagehelper/pagehelper
      compile group: 'com.github.pagehelper', name: 'pagehelper', version: '5.1.2'
      // https://mvnrepository.com/artifact/org.aspectj/aspectjweaver
      compile group: 'org.aspectj', name: 'aspectjweaver', version: '1.9.2'
      MybatisGenerator 'org.Mybatis.generator:Mybatis-generator-core:1.3.2'
      MybatisGenerator 'mysql:mysql-connector-java:5.1.38'
      MybatisGenerator 'tk.Mybatis:mapper:3.3.1'
   }
   def getDbProperties = {
   def properties = new Properties()
      file("src/main/resources/Mybatis/db-mysql.properties").withInputStream
{ inputStream ->
          properties.load(inputStream)
      }
      properties;
   }
   task MybatisGenerate << {
      def properties = getDbProperties()
      ant.properties['targetProject'] = projectDir.path
      ant.properties['driverClass'] = properties.getProperty("jdbc.driverClassName")
      ant.properties['connectionURL'] = properties.getProperty("jdbc.url")
```

```
    ant.properties['userId'] = properties.getProperty("jdbc.user")
    ant.properties['password'] = properties.getProperty("jdbc.pass")
    ant.properties['src_main_java'] = sourceSets.main.java.srcDirs[0].path
    ant.properties['src_main_resources'] = sourceSets.main.resources.srcDirs[0].path
    ant.properties['modelPackage'] = properties.getProperty("package.model")
    ant.properties['mapperPackage'] = properties.getProperty("package.mapper")
    ant.properties['sqlMapperPackage'] =properties.getProperty("package.xml")
    ant.taskdef(
           name: 'mbgenerator',
    classname: 'org.Mybatis.generator.ant.GeneratorAntTask',
     classpath: configurations.MybatisGenerator.asPath)
    ant.mbgenerator(overwrite: true,
     configfile: 'src/main/resources/Mybatis/generatorConfig.xml', verbose: true)
{
        propertyset {
            propertyref(name: 'targetProject')
            propertyref(name: 'userId')
            propertyref(name: 'driverClass')
            propertyref(name: 'connectionURL')
            propertyref(name: 'password')
            propertyref(name: 'src_main_java')
            propertyref(name: 'src_main_resources')
            propertyref(name: 'modelPackage')
            propertyref(name: 'mapperPackage')
            propertyref(name: 'sqlMapperPackage')
        }
    }
}
```

2．webapp/WEB-INF/web.xml 文件

webapp/WEB-INF/web.xml 文件内容如下：

```
<?xml version="1.0" encoding="UTF-8"?>
<web-app version="2.5" xmlns="http://java.sun.com/xml/ns/javaee"
   xmlns:xsi="http://www.w3.org/2001/XMLSchema-instance"
   xsi:schemaLocation="http://java.sun.com/xml/ns/javaee
   http://java.sun.com/xml/ns/javaee/web-app_2_5.xsd">
   <!-- 配置 Spring 的监听 -->
   <!--配置启动 IoC 容器的 Listener-->
   <context-param>
      <param-name>contextConfigLocation</param-name>
      <param-value>classpath*:spring-mybatis.xml</param-value>
   </context-param>
   <!--配置 Spring listener-->
   <listener>
    <listener-class>org.springframework.web.context.ContextLoaderListener
</listener-class>
   </listener>
   <!--解决 POST 乱码问题-->
```

```
    <filter>
       <filter-name>CharacterEncodingFilter</filter-name>
       <filter-class>org.springframework.web.filter.CharacterEncodingFilter
</filter-class>
       <init-param>
          <param-name>encoding</param-name>
          <param-value>utf-8</param-value>
       </init-param>
    </filter>
    <filter-mapping>
       <filter-name>CharacterEncodingFilter</filter-name>
       <url-pattern>*.do</url-pattern>
    </filter-mapping>
    <!--Springmvc 前端控制器配置-->
    <servlet>
       <servlet-name>dispatcherServlet</servlet-name>
       <servlet-class>org.springframework.web.servlet.DispatcherServlet</servlet-
class>
       <init-param>
          <param-name>contextConfigLocation</param-name>
          <param-value>classpath:Spring-mvc.xml</param-value>
       </init-param>
       <load-on-startup>1</load-on-startup>
       <multipart-config>
          <!--上传到 d:/upload 目录-->
          <location>d:/upload</location>
          <!--文件大小为 2MB-->
          <max-file-size>2097152</max-file-size>
          <!--整个请求不超过 4MB-->
          <max-request-size>4194304</max-request-size>
          <!--所有文件都要写入磁盘-->
         <file-size-threshold>0</file-size-threshold>
       </multipart-config>
    </servlet>
    <servlet-mapping>
       <servlet-name>dispatcherServlet</servlet-name>
       <url-pattern>/</url-pattern>
    </servlet-mapping>
 </web-app>
```

3. resources/spring-mvc.xml 文件

resources/spring-mvc.xml 文件内容如下：

```
<?xml version="1.0" encoding="UTF-8"?>
<beans xmlns="http://www.springframework.org/schema/beans"
      xmlns:xsi="http://www.w3.org/2001/XMLSchema-instance"
      xmlns:context="http://www.springframework.org/schema/context"
      xmlns:mvc="http://www.Springframework.org/schema/mvc"
      xsi:schemaLocation="http://www.springframework.org/schema/beans
```

```
http://www.springframework.org/schema/beans/spring-beans.xsd
http://www.springframework.org/schema/context
http://www.springframework.org/schema/context/spring-context.xsd
http://www.springframework.org/schema/mvc
http://www.springframework.org/schema/mvc/spring-mvc.xsd">
        <!-- 配置自动扫描的包 --> <!-- 自动扫描控制器 -->
        <context:component-scan base-package="com.ssm"/>
        <!-- 视图渲染 -->
        <bean id="internalResourceViewResolver"
              class="org.springframework.web.servlet.view.InternalResourceViewResolver">
            <property name="prefix" value="/WEB-INF/views/"/>
            <property name="suffix" value=".jsp"/>
        </bean>
        <!-- 校验器，配置 validator -->
        <bean  id="validator"  class="org.springframework.validation.beanvalidation.
LocalValidatorFactoryBean">
            <property  name="providerClass"  value="org.hibernate.validator.Hibernate
Validator"></property>
        </bean>
        <bean id="multipartResolver" class="org.springframework.web.multipart.support.
StandardServletMultipartResolver"/>
        <bean  id="simpleMappingExceptionResolver"  class="org.springframework.web.
servlet.handler.simpleMappingExceptionResolver">
        <property name="exceptionMappings">
            <map>
                <!-- key:异常类别（非全称），视图名称 -->
                <entry key="DatabaseException" value="databaseError"/>
                <entry key="InvalidCreditCardException" value="creditCardError"/>
            </map>
        </property>
        <!-- 默认的错误处理页面，异常的名称 -->
        <property name="defaultErrorView" value="error"/>
        <property name="exceptionAttribute" value="ex"/>
    </bean>
        <!-- 控制器映射器和控制器适配器 -->
        <mvc:annotation-driven  validator="validator"></mvc:annotation-driven>
        <!-- 控制器映射器和控制器适配器 -->
        <mvc:annotation-driven></mvc:annotation-driven>
        <!-- 静态资源映射器 -->
        <mvc:resources mapping="/views/**" location="/WEB-INF/views/" />
        <mvc:resources mapping="bootstrap/**" location="bootstrap/" />
        <mvc:interceptors>
            <mvc:interceptor>
                <mvc:mapping path="/Admin/*"/>
                <bean class="com.ssm.inter.MyInterceptor"/>
            </mvc:interceptor>
        </mvc:interceptors>
    </beans>
```

4．resources/spring-mybatis.xml 文件

resources/spring-mybatis.xml 文件内容如下：

```
<?xml version="1.0" encoding="UTF-8"?>
<beans xmlns="http://www.springframework.org/schema/beans"
      xmlns:xsi="http://www.w3.org/2001/XMLSchema-instance"
      xmlns:context="http://www.springframework.org/schema/context"
      xmlns:tx="http://www.springframework.org/schema/tx" xmlns:aop="http://www.
springframework.org/schema/aop"
      xsi:schemaLocation="http://www.springframework.org/schema/beans
      http://www.springframework.org/schema/beans/spring-beans.xsd
      http://www.springframework.org/schema/context
      http://www.springframework.org/schema/context/spring-context.xsd
      http://www.springframework.org/schema/tx
      http://www.springframework.org/schema/tx/spring-tx.xsd   http://www.spring
framework.org/schema/aop http://www.springframework.org/schema/aop/spring-aop.xsd">
    <!-- 配置数据库相关参数properties的属性：${url} -->
    <context:property-placeholder location="classpath*:jdbc.properties"/>
    <!-- 数据库连接池 -->
    <bean   id="dataSource"    class="com.mchange.v2.c3p0.ComboPooledDataSource"
destroy-method="close">
        <property name="driverClass" value="${jdbc.driver}"/>
        <property name="jdbcUrl" value="${jdbc.url}"/>
        <property name="user" value="${jdbc.username}"/>
        <property name="password" value="${jdbc.password}"/>
    </bean>
    <!-- Spring和MyBatis完美整合，不需要MyBatis的配置映射文件 -->
    <bean id="sqlSessionFactory" class="org.mybatis.spring.SqlSessionFactoryBean">
        <property name="dataSource" ref="dataSource" />
        <!-- 自动扫描mapping.xml文件 -->
        <property    name="mapperLocations"    value="classpath*:/mapper/*.xml">
</property>
        <!-- 配置分页插件 -->
        <property name="plugins">
            <array>
                <bean class="com.github.pagehelper.PageInterceptor">
                    <property name="properties">
                        <!--使用下面的方式配置参数，一行配置一个 -->
                        <value>
                            helperDialect=mysql
                        </value>
                    </property>
                </bean>
            </array>
        </property>
    </bean>
    <!-- DAO接口所在包名，Spring会自动查找其下的类 -->
    <bean class="org.mybatis.spring.mapper.MapperScannerConfigurer">
```

```
        <property name="basePackage" value="com.ssm.dao" />
        <property   name="sqlSessionFactoryBeanName"   value="sqlSessionFactory">
</property>
     </bean>
     <bean id="sqlSessionTemplate" class="org.mybatis.spring.SqlSessionTemplate">
        <constructor-arg index="0" ref="sqlSessionFactory" />
     </bean>
     <!--(事务管理)transactionManager,使用 DataSourceTransactionManager 类创建的 Bean,
注入 Bean-dataSource -->
     <bean id="transactionManager"
          class="org.springframework.jdbc.datasource.DataSourceTransaction
Manager">
        <property name="dataSource" ref="dataSource" />
     </bean>
     <!--定义增强，使用前面的事务管理 Bean-transactionManager-->
     <tx:advice id="txAdvice" transaction-manager="transactionManager">
        <tx:attributes>
           <!--定义 get 字符开头的方法，只读事务-->
           <tx:method name="get*" read-only="true"/>
           <!--定义 insert 字符开头的方法，遇到 Exception 回滚-->
           <tx:method name="insert*" rollback-for="Exception"/>
           <!--定义 update*,delete*方法-->
           <tx:method name="update*,delete*"/>
        </tx:attributes>   </tx:advice>
     <!--使用 Schema 方式定义切面，切点是 com.ssm.service.impl 包下所有类的所有方法，增强是前
面定义的 txAdvice-->
     <aop:config>
        <aop:pointcut id="serviceOperation" expression= "execution(* com.ssm.service.
impl.*.*(..))" />
        <aop:advisor advice-ref="txAdvice" pointcut-ref="serviceOperation" />
     </aop:config>
  </beans>
```

5. jdbc.properties 属性文件

jdbc.properties 属性文件内容如下：

```
jdbc.driver=com.mysql.jdbc.Driver
jdbc.url=jdbc:mysql://localhost:3306/test?useUnicode=true&characterEncoding=UTF-8
jdbc.username=root
jdbc.password=123
#定义初始连接数
initialSize=0
#定义最大连接数
maxActive=20
#定义最大空闲
maxIdle=20
#定义最小空闲
minIdle=1
```

```
#定义最长等待时间 1 分钟
maxWait=60000
```

小　结

AOP 通过横向切割的方式抽取分散在业务代码中的相同代码，构成一个独立的模块，还业务逻辑类一个清晰的世界。通过 OOP 的纵向抽象和 AOP 的横向抽取，真正解决了大型项目共有的重复性代码问题。Spring AOP 用非容器依赖的编程方式实现了与其他框架的无侵入的无缝结合，并且降低了其他框架的使用难度。它借用 Spring IoC 核心和反射代理机制及 AspectJ 框架对面向切面编程进行了完美的诠释和实现，实现了功能代码的模块化，大大减轻了开发人员的工作量。Spring AOP 简洁方便、丰富的配置形式，使得开发人员可以集中精力进行面向切面的编程。基于 Spring AOP 的声明式事务管理，使得程序员不需要手动编码去维护事务。总而言之，Spring AOP 大大提升了项目模块化开发的效率和质量。

习　题

一、单选题

1．AOP 技术的优势在于（　　）。

A．通过接口、抽象及组合增强对象内部能力

B．将核心关注点与横切关注点完全隔离

C．有利于增强业务的安全性

D．让一些公共逻辑更好地分布在对象核心逻辑中

2．Spring 中 AOP 术语的全称是（　　）。

A．依赖注入　　B．切面编程　　C．面向对象编程　　D．面向切面编程

二、多选题

1．AOP 的基本概念有（　　）。

A．Advice：用于定义拦截行为

B．Join Point：提供访问当前被通知方法的目标对象、代理对象、方法参数等数据

C．Pointcut：捕获所有的连接点在指定的方法中执行，包括执行方法本身

D．Aspect：切点指示符，用来指示切点表达式的目标对象

2．Spring 的 AOP 的动态代理机制有（　　）。

A．CGLIB 库　　B．JDK 依赖注入　　C．Proxy 代理　　D．Auto 代理

三、填空题

1．AOP，也就是________，应运而生，专门解决一些具有横切性质的系统性服务，如事务管理、安全检查、缓存管理、对象池管理等。

2．Spring 支持 5 种类型的增强：________、________、________、________、________。

3．Spring________是 Spring AOP 技术的精彩应用，它作为一个切面织入目标业务方法，

使得业务代码从事务代码中脱离出来。

4．动态代理实现有两种方式：________和________。

综合实训

实训 1．设计一个环绕增强，并检验。

实训 2．在项目中设计一个 Advice 类，用于判断用户是否具备增加学生的权限（该权限属于管理员身份）。

如果不是管理员，单击“增加学生”按钮，弹出报错对话框。

实训 3．研究 Spring 权限框架，看能否加入基于 Gradle 构建的 Module 中。

第8章

项目快速开发框架 Spring Boot

本章学习目标

- 了解 Spring Boot 的原理
- 掌握 Spring Boot 的开发步骤
- 掌握 Maven 构建工具的配置和管理

本章将简单介绍 Spring Boot 框架的特性和原理，重点讲述如何使用 Spring Boot 完成项目中三大框架的集成和部署运行。

8.1 Spring Boot

Spring Boot 是 Pivotal 团队开发的框架，其设计初衷是简化 Spring Web 应用的搭建以加快整个开发过程。该框架使用了特定的方式来配置项目，使得开发人员不再需要定义复杂且重复的配置。凭借这个优势，在发布速度比技术完美更为重要的快速应用开发领域，Spring Boot 成为重要的弄潮儿。

总结起来，Spring Boot 可以帮助我们简单、快速、方便地搭建项目，对主流开发框架可以实现无配置集成，极大提高了开发和部署效率。

8.1.1 Spring Boot 的原理和特点

从本质上来说，Spring Boot 就是一些库的集合，它能够被任意项目构建工具所使用。Spring Boot 提供了命令行界面（CLI），它可以用来运行和测试 Spring Boot 应用。Spring Boot 及其 CLI 可以在 Spring 仓库中手动下载和安装，更为简便的方式是使用 Groovy 环境管理器（GVM）如 IDEA 平台进行 Boot 版本的安装和管理。

对使用 Spring Boot 的 Web 项目来说，如果要打包和分发，需要依赖于 Maven 或 Gradle 这样的构建工具。Spring Boot 的功能是模块化的，通过导入 starter initializer 模块，可以将许多依赖一次性添加到项目之中，简化了项目的依赖结构。为了更容易地管理依赖版本和使用默认配置，Spring Boot 还提供了一个 parent POM，通过继承它可简化项目开发。

总结起来，Spring Boot 框架具备以下特征：

（1）可独立创建基于 Maven 或 Gradle 构建工具的 Spring 应用程序，并且可以创建可执行的 jars 包和 Web 项目。

（2）内嵌 Tomcat 或 Jetty 等 Web 容器。

（3）提供自动配置的 starter 项目对象模型（POMS）以简化 Maven 配置。

（4）自动配置 Spring IoC 容器。

（5）提供准备好的特性，如指标、健康检查和外部化配置。

（6）不生成代码，不需要 XML 配置。

8.1.2　任务一：Spring Boot 快速开发

【示例 8-1】创建 Spring Boot 的 Module（模块），完成集成和数据库连接配置，以及数据库持久层的自动生成。最后，显示 student 表中所有学生的信息。

步骤 1．创建基于 Maven 的 Spring Boot 项目。

如图 8-1 所示，创建 Spring initializr 的 Module（模块），在 Module SDK 下拉列表里选择默认的 Project SDK（1.8）。

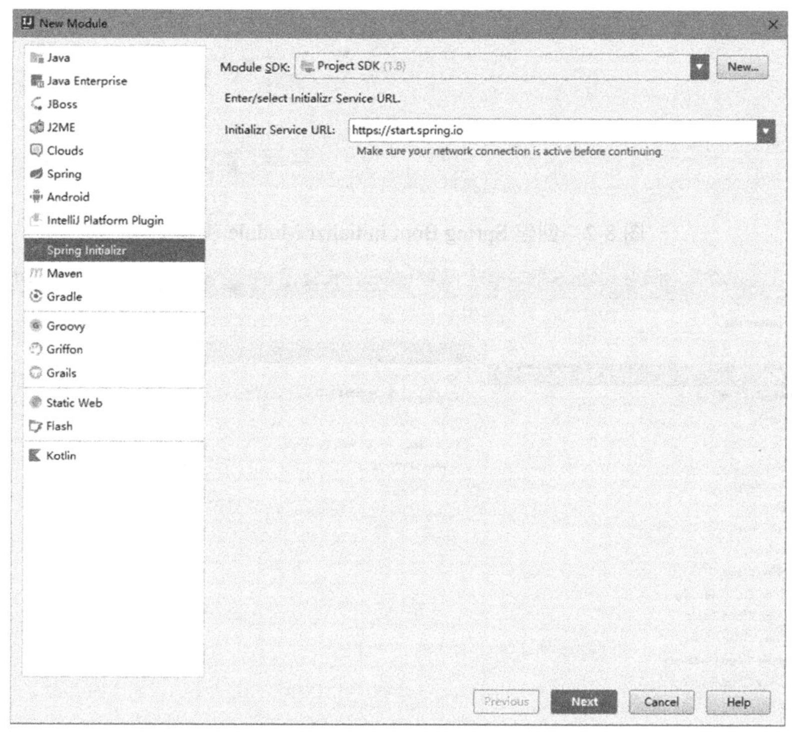

图 8-1　创建 Spring Boot initializr Module 步骤 1

单击 Next 按钮，出现如图 8-2 所示界面，在 Group 输入框填入组织名，在 Artifact 输入框填入项目成品的名称，在 Type 下拉列表里选择 Maven Project，表明使用 Maven 来构建 Module。在 8.2 节中我们会详细介绍 Maven 的安装配置。

单击 Next 按钮，出现如图 8-3 所示界面，在 Dependencies 左侧列表框中选择 Web，中间列表框中勾选 Spring Web 复选框和 Spring Web Services 复选框。然后在左侧列表框中选择 SQL，在中间列表框中勾选 JDBC API 复选框、MyBatis Framework 复选框和 MySQL Driver 复选框，如图 8-4 所示。

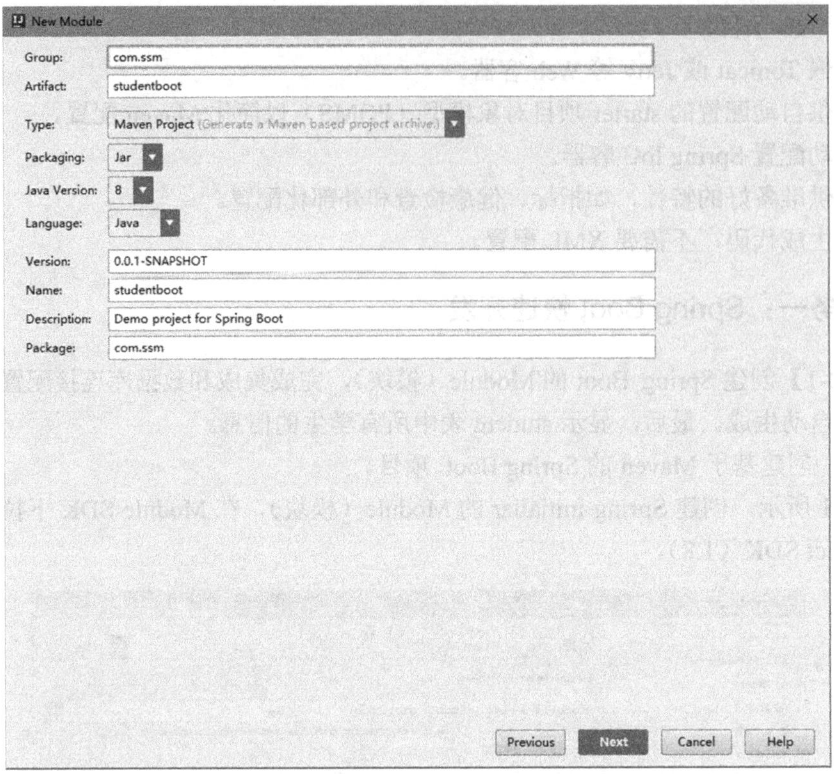

New Module

Group: com.ssm
Artifact: studentboot
Type: Maven Project (Generate a Maven based project archive)
Packaging: Jar
Java Version: 8
Language: Java
Version: 0.0.1-SNAPSHOT
Name: studentboot
Description: Demo project for Spring Boot
Package: com.ssm

Previous Next Cancel Help

图 8-2　创建 Spring Boot initializr Module 步骤 2

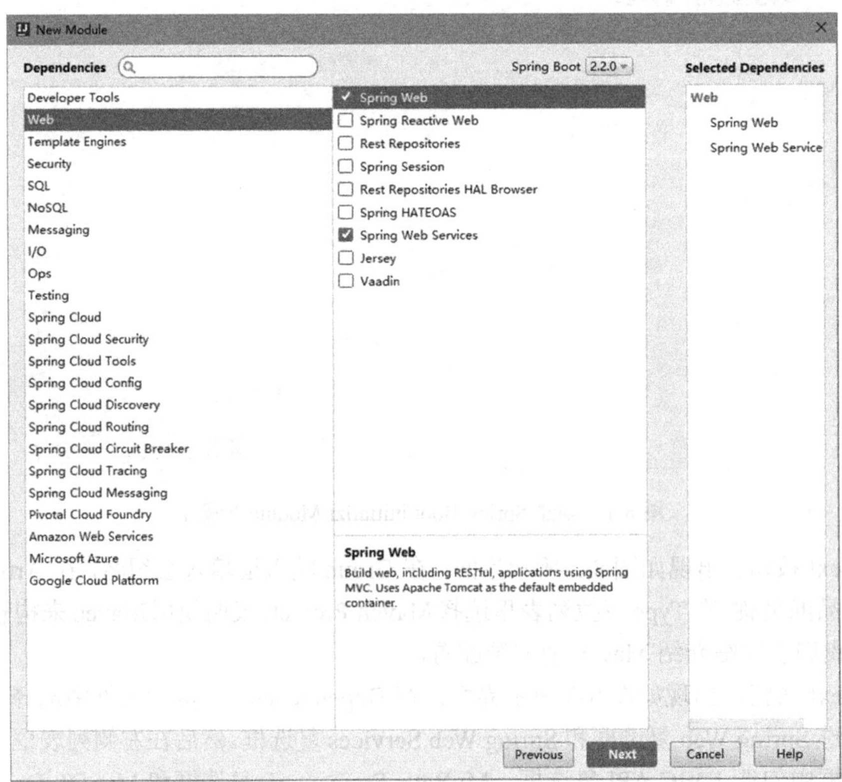

图 8-3　创建 Spring Boot initializr Module 步骤 3

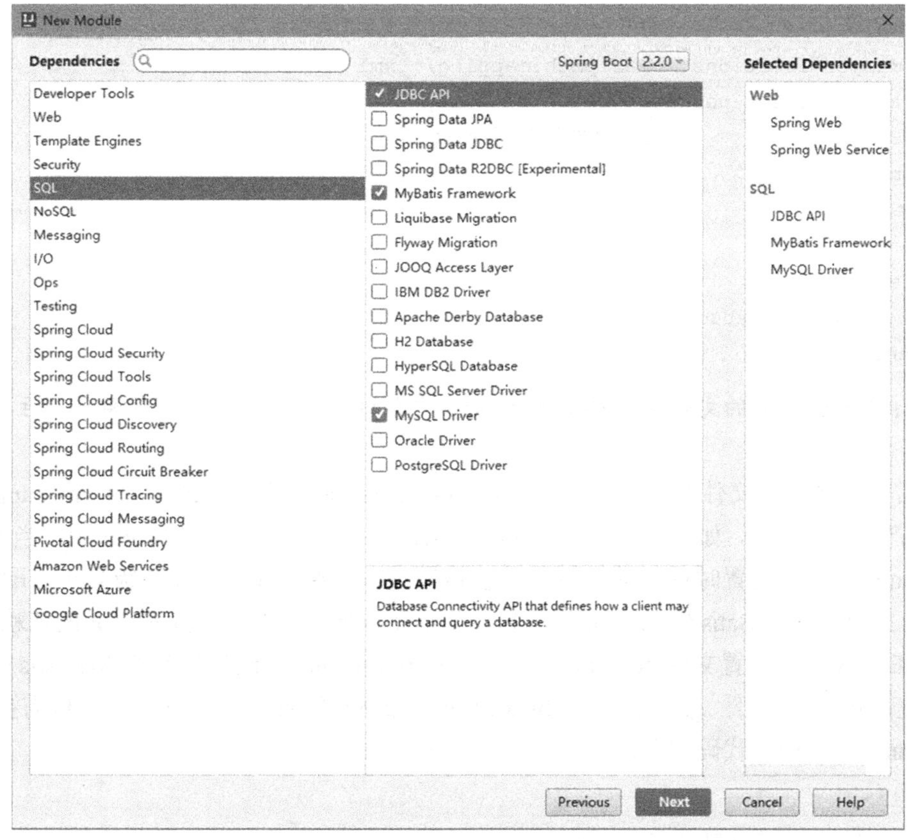

图 8-4 创建 Spring Boot initializr Module 步骤 4

然后单击 Next 按钮，接着单击 Finish 按钮，等待项目构建结束。

步骤 2. 完善项目配置的属性文件。

删去 resources 文件夹下的 application.properties 属性文件，新建 Spring 框架的配置文件 application.yml，在文件里定义端口、数据源及 Spring MVC 的属性等。具体代码如下：

```
server:
    port: 8080          //服务器端口号
Spring:
  datasource:           //数据源信息：驱动器、连接字符串、用户名、密码
      username: root
      password: 123
      url:
  jdbc:mysql://localhost:3306/test?useUnicode=true&characterEncoding=ut  f-8&
useSSL=true&serverTimezone=UTC
      driver-class-name: com.mysql.jdbc.Driver
  profiles:
      active: dev
  mvc:
   view:                //静态资源文件所在位置和文件扩展名
       prefix:  /WEB-INF/views/
       suffix:  .jsp
```

```
Mybatis:                    //定义 myBatis 框架的映射文件所在位置
      mapper-locations: classpath:mapping/*.xml
      type-aliases-package: com.ssm.entity
#showSql                    //定义调试状态下显示 SQL 语句
logging:
  level:
    com:
      example:
        mapper : debug
    web: debug
```

提示：YML 格式的文件使用缩进表示 key 之间的父子关系，书写时要特别注意缩进和对齐。

YML 格式的配置文件有 application.yml、application-dev（开发测试环境）.yml、application-prod（生产环境）.yml。如果激活 application-dev.yml 或者 application-prod.yml，当同时存在 application.yml 同名配置时后者的配置属性会被覆盖（激活配置文件优先级高于总配置文件）。

步骤 3．使用 MyBatis Generator 工具自动生成持久层，添加数据库和 JSP 的依赖。

（1）在 Maven 的配置文件 pom.xml 里添加 MyBatis Generator 插件。定义 MyBatis Generator 运行时使用的配置文件 generatorConfig.xml 所在的位置 src/main/resources/和需要的依赖 mysql-connector-java。代码如下：

```
<plugin>
   <groupId>org.Mybatis.generator</groupId>
<artifactId>Mybatis-generator-maven-plugin</artifactId>
     <version>1.3.2</version>
      <executions>
       <execution>
          <id>Generate Mybatis Artifacts</id>
          <phase>deploy</phase>
          <goals>
             <goal>generate</goal>
          </goals>
       </execution>
    </executions>
      <configuration>
       <!-- generator 工具配置文件的位置 --> <configurationFile>src/main/resources/
generatorConfig.xml</configurationFile>
       <verbose>true</verbose>
       <overwrite>true</overwrite>
    </configuration>
    <dependencies>
       <dependency>
          <groupId>mysql</groupId>
<artifactId>mysql-connector-java</artifactId>
          <version>5.1.34</version>
```

```
        </dependency>
        <dependency>
           <groupId>org.Mybatis.generator</groupId>
    <artifactId>Mybatis-generator-core</artifactId>
           <version>1.3.2</version>
        </dependency>
      </dependencies>
    </plugin>
```

(2)在 pom.xml 文件里添加依赖，分别是 MySQL 驱动、SpringBoot test 框架集、内嵌 tomcat、jstl、servlet-api、pagehelper 分页插件。代码如下：

```
    <dependency>
      <groupId>mysql</groupId>
      <artifactId>mysql-connector-java</artifactId>
      <scope>runtime</scope>
       </dependency>
       <dependency>
      <groupId>org.springframework.boot</groupId>
<artifactId>spring-boot-starter-test</artifactId>
      <scope>test</scope>
      <exclusions>
        <exclusion>
          <groupId>org.junit.vintage</groupId>
<artifactId>junit-vintage-engine</artifactId>
        </exclusion>
         </exclusions>
       </dependency>
       <dependency>
      <groupId>org.apache.tomcat.embed</groupId>
      <artifactId>tomcat-embed-jasper</artifactId>
      <version>8.5.20</version>
       </dependency>
       <dependency>
      <groupId>javax.servlet</groupId>
      <artifactId>jstl</artifactId>
      <version>1.2</version>
       </dependency>
       <dependency>
      <groupId>javax.servlet</groupId>
      <artifactId>javax.servlet-api</artifactId>
      <version>3.1.0</version>
       </dependency>
       <!-- https://mvnrepository.com/artifact/com.github.pagehelper/pagehelper- Spring-
boot-starter -->
       <dependency>
      <groupId>com.github.pagehelper</groupId>
```

```
<artifactId>pagehelper-Spring-boot-starter</artifactId>
      <version>1.2.10</version>
      </dependency>
```

步骤 4. 创建 MyBatis Generator 需要的 generatorConfig.xml 文件。在 resources 文件夹下新建 generatorConfig.xml，生成注解形式的 DAO，元素的具体功能参考文件中的注释。代码如下：

```
<?xml version="1.0" encoding="UTF-8"?>
<!DOCTYPE generatorConfiguration
        PUBLIC "-//Mybatis.org//DTD Mybatis Generator Configuration 1.0//EN"
        "http://Mybatis.org/dtd/Mybatis-generator-config_1_0.dtd">
<!-- 配置生成器 -->
<generatorConfiguration>
    <!--执行 generator 插件生成文件的命令： call mvn Mybatis-generator:generate -e -->
    <!-- 引入配置文件 -->
    <properties resource="MybatisGenerator.properties"/>
        <!-- 一个数据库一个 context -->
    <!--defaultModelType="flat" 大数据字段，不分表 -->
    <context id="MysqlTables" targetRuntime="Mybatis3Simple" defaultModelType=
"flat">
        <!-- 自动识别数据库关键字，默认 false，如果设置为 true，根据 SqlReservedWords 中定义
关键字列表；一般保留默认值，遇到数据库关键字（Java 关键字），使用 columnOverride 覆盖 -->
        <property name="autoDelimitKeywords" value="true" />
        <!-- 生成的 Java 文件的编码 -->
        <property name="javaFileEncoding" value="utf-8" />
        <!-- beginningDelimiter 和 endingDelimiter：指明数据库的用于标记数据库对象名的符号，
比如 Oracle 就是双引号，MySQL 默认是`反引号； -->
        <property name="beginningDelimiter" value="`" />
        <property name="endingDelimiter" value="`" />
        <!-- 格式化 java 代码 -->
        <property  name="javaFormatter"  value="org.Mybatis.generator.api.dom.
DefaultJavaFormatter"/>
        <!-- 格式化 XML 代码 -->
        <property  name="xmlFormatter"  value="org.Mybatis.generator.api.dom.
DefaultXmlFormatter"/>
        <plugin type="org.Mybatis.generator.plugins.SerializablePlugin" />
        <plugin type="org.Mybatis.generator.plugins.ToStringPlugin" />
        <!-- 注释 -->
        <commentGenerator >
            <property name="suppressAllComments" value="false"/><!-- 是否取消注释 -->
            <property name="suppressDate" value="true" /> <!-- 是否生成注释代时间戳-->
        </commentGenerator>
        <!-- jdbc 连接 -->
        <jdbcConnection  driverClass="${jdbc_driver}"  connectionURL="${jdbc_url}"
userId="${jdbc_user}" password="${jdbc_password}" />
        <!-- 类型转换 -->
```

```
        <javaTypeResolver>
            <!-- 是否使用 bigDecimal, false 可自动转化以下类型(Long, Integer, Short, etc.) -->
            <property name="forceBigDecimals" value="false"/>
        </javaTypeResolver>
        <!-- 生成实体类地址 -->
        <javaModelGenerator targetPackage="com.ssm.entity" targetProject= "${project}" >
            <property name="enableSubPackages" value="false"/>
            <property name="trimStrings" value="true"/>
        </javaModelGenerator>
        <!-- 生成 mapxml 文件 -->
        <sqlMapGenerator  targetPackage="mappers" targetProject="${resources}" >
            <property name="enableSubPackages" value="false" />
        </sqlMapGenerator>
        <!-- type="ANNOTATEDMAPPER" 表示生成注解形式 sql 的 dao-->
        <javaClientGenerator  type="ANNOTATEDMAPPER"  targetPackage="com.ssm.dao"
targetProject="${project}" >
            <property name="enableSubPackages" value="false" />
        </javaClientGenerator>
        <!-- table 可以有多个,每个数据库中的表都可以写一个 table, tableName 表示要匹配的数据
库表,也可以在 tableName 属性中通过使用%通配符来匹配所有数据库表,只有匹配的表才会自动生成文件 -->
        <table  tableName="%"  enableCountByExample="true"  enableUpdateByExample=
"true"  enableDeleteByExample="true"  enableSelectByExample="true"  selectByExample
QueryId="true">
            <property name="useActualColumnNames" value="false" />
            <!-- 数据库表主键 -->
            <generatedKey column="id" sqlStatement="Mysql" identity="true" />
        </table>
    </context>
</generatorConfiguration>
```

步骤 5. 在 resources 文件夹下新建 mybatisGenerator.properties 文件，此文件用于 MyBatis Generator 自动生成持久层时的数据库连接。代码如下：

```
#Mybatis Generator configuration
#dao 类和实体类的位置
project =src/main/java
#mapper 文件的位置
resources=src/main/resources
#根据数据库中的表生成对应的 pojo 类、dao、mapper
jdbc_driver =com.mysql.jdbc.Driver
jdbc_url=jdbc:mysql://localhost:3306/test
jdbc_user=root
jdbc_password=123
```

步骤 6. 运行 MyBatis Generator 插件，生成持久层代码。

单击右侧边栏的 Maven Projects，刷新 Maven 项目，双击 Plugins→mybatis-generator→mybatis-generator:generate，如图 8-5 所示。

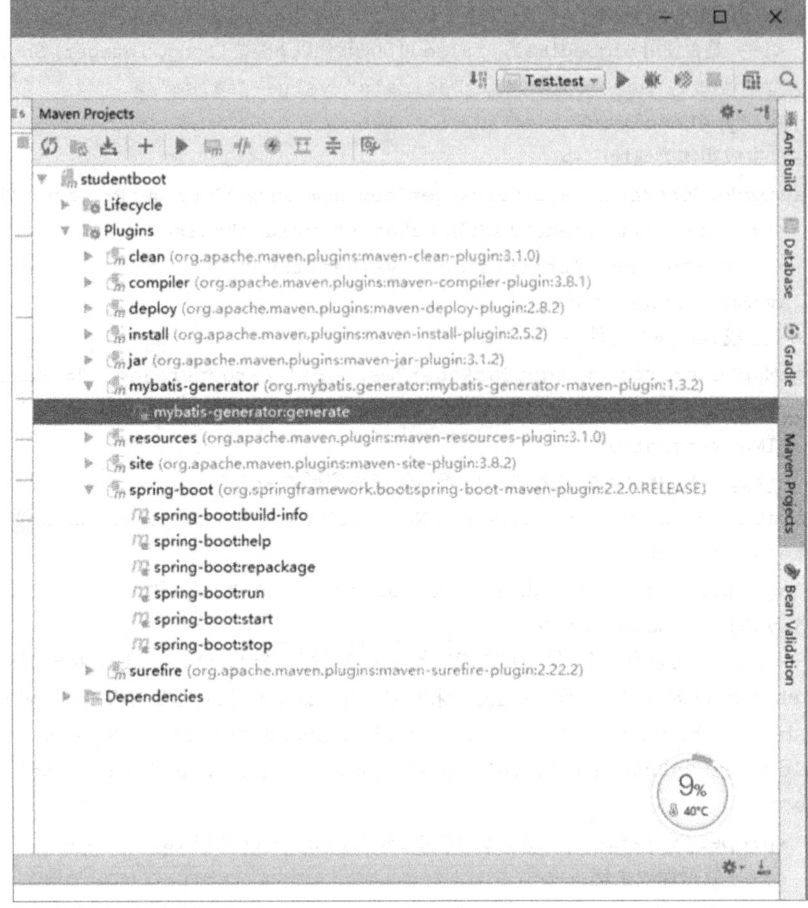

图 8-5　运行 MyBatis Generator 插件

控制台如果输出如下信息，则表示自动生成成功了。检查左边的 Project→studentboot 的树状目录里，是否有 com.ssm.dao 和 com.ssm.entity 两个文件夹及里面的文件。

```
[INFO] --- Mybatis-generator-maven-plugin:1.3.2:generate (default-cli) @ studentboot ---
[INFO] Connecting to the Database
[INFO] Introspecting table %
[INFO] Generating Record class for table student
[INFO] Generating Mapper Interface for table student
[INFO] Generating Record class for table score
[INFO] Generating Mapper Interface for table score
[INFO] Generating Record class for table admin
[INFO] Generating Mapper Interface for table admin
[INFO] Generating Record class for table course
[INFO] Generating Mapper Interface for table course
[INFO] Generating Record class for table teacher
[INFO] Generating Mapper Interface for table teacher
[INFO] Saving file Student.java
[INFO] Saving file StudentMapper.java
[INFO] Saving file Score.java
```

```
[INFO] Saving file ScoreMapper.java
[INFO] Saving file Admin.java
[INFO] Saving file AdminMapper.java
[INFO] Saving file Course.java
[INFO] Saving file CourseMapper.java
[INFO] Saving file Teacher.java
[INFO] Saving file TeacherMapper.java
[INFO] ------------------------------------------------------------------------
[INFO] BUILD SUCCESS
[INFO] ------------------------------------------------------------------------
[INFO] Total time: 1.521 s
[INFO] Finished at: 2019-10-20T12:26:41+08:00
[INFO] Final Memory: 15M/245M
[INFO] ------------------------------------------------------------------------
```

步骤 7. 编写业务程序，完成 controller 和 service 包下的源文件，实现 student 表中数据在网页上的显示。

在 controller 包下编写文件 StudentController.java，在里面定义一个 list.do，内容见源码。在 service 包下编写接口文件 StudentService.java，并进行 list 方法的声明。在 service/impl 包下新建 StudentServiceImpl.java，实现接口文件 StudentService.java 中的 list 方法。

修改 dao 包下的 StudentMapper.java 文件，添加如下内容，实现 list 的 DAO 层方法。

```
    @Select({
            "select",
            "* ",
            "from student"
     })
     @Results({
            @Result(column="Id", property="id", jdbcType=JdbcType.INTEGER, id=true),
            @Result(column="name", property="name", jdbcType=JdbcType.VARCHAR),
            @Result(column="password",   property="password",   jdbcType=JdbcType.
VARCHAR),
            @Result(column="sex", property="sex", jdbcType=JdbcType.INTEGER),
            @Result(column="clazz", property="clazz", jdbcType=JdbcType.VARCHAR),
            @Result(column="birthday",   property="birthday",   jdbcType=JdbcType.
VARCHAR),
            @Result(column="image", property="image", jdbcType=JdbcType.VARCHAR),
     })
     List<Student> list();
```

步骤 8. 添加对 dao 包的自动扫描。

在 StudentbootApplicaiton.java 文件中添加@MapperScan("com.ssm.dao")注解，表示扫描的 mapper 映射位于 com.ssm.dao 包下，该注解放在@SpringBootApplication 注解之后。

步骤 9. 创建 webapp 目录，Bootstrap 框架的源文件和 listStudents.jsp 文件可以从示例 5-1 的源码中复制过来。最终的目录结构如图 8-6 所示。

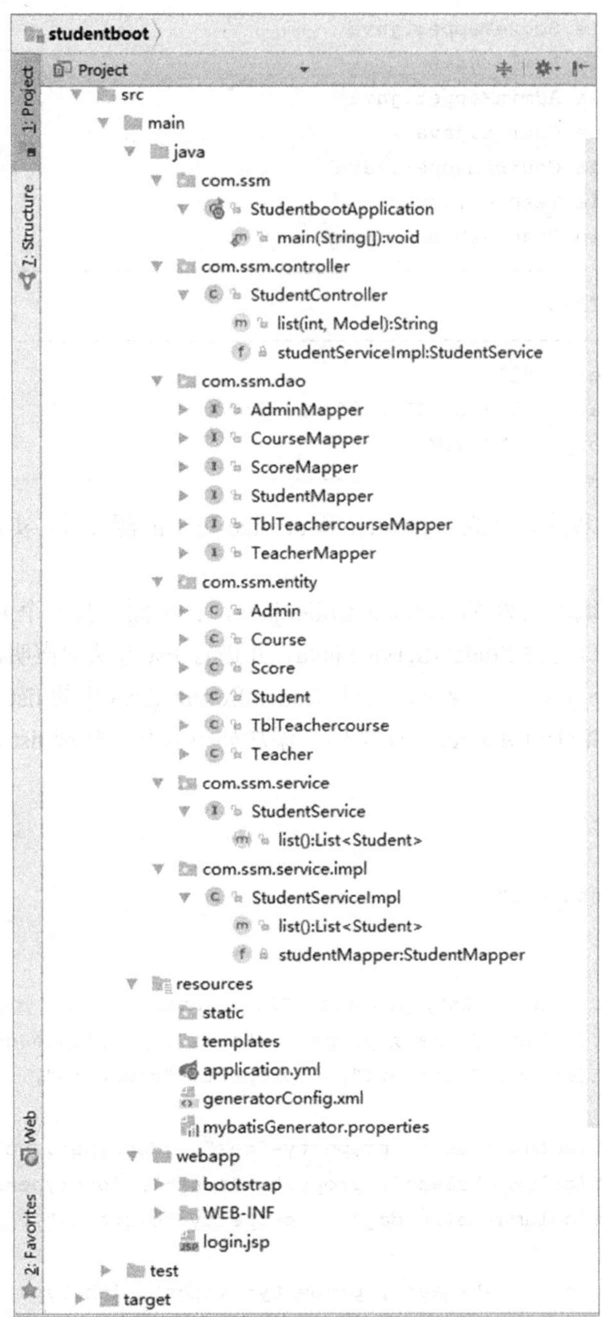

图 8-6　studentboot 项目的目录结构

步骤 10．运行项目，控制台检查 Web 项目的运行结果。

单击右侧边栏 Maven Projects，展开 Plugins→spring-boot，双击 spring-boot:run，控制台显示如下信息，则表示运行成功。

```
  2019-10-21  09:56:52.223  INFO  20412  ---  [                    main]
o.s.s.concurrent.ThreadPoolTaskExecutor     :  Initializing  ExecutorService
'applicationTaskExecutor'
  2019-10-21  09:56:52.223  DEBUG  20412  ---  [                   main]
```

```
s.w.s.m.m.a.RequestMappingHandlerAdapter : ControllerAdvice beans: 0 @ModelAttribute, 0 @InitBinder, 1 RequestBodyAdvice, 1 ResponseBodyAdvice
  2019-10-21 09:56:52.254 DEBUG 20412 --- [ main] s.w.s.m.m.a.RequestMappingHandlerMapping : 3 mappings in 'requestMappingHandlerMapping'
  2019-10-21 09:56:52.285 DEBUG 20412 --- [ main] o.s.w.s.handler.SimpleUrlHandlerMapping : Patterns [/webjars/**, /**] in 'resourceHandlerMapping'
  2019-10-21 09:56:52.285 DEBUG 20412 --- [ main] .m.m.a.ExceptionHandlerExceptionResolver : ControllerAdvice beans: 0 @ExceptionHandler, 1 ResponseBodyAdvice
  2019-10-21 09:56:52.410 INFO 20412 --- [ main] o.s.b.w.embedded.tomcat.TomcatWebServer : Tomcat started on port(s): 8080 (http) with context path ''
  2019-10-21 09:56:52.410 INFO 20412 --- [ main] com.ssm.StudentbootApplication : Started StudentbootApplication in 1.95 seconds (JVM running for 2.36)
```

步骤 11．服务器启动完毕后，在浏览器地址栏输入 http://localhost:8080/Student/list.do，运行结果如图 8-7 所示。

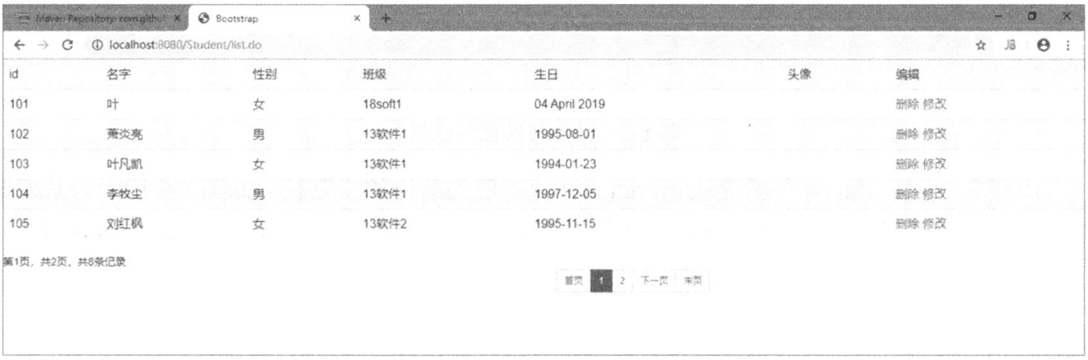

图 8-7　运行结果

示例 8-1 演示了 Spring Boot 项目的创建、配置、编程、部署及运行访问。我们会发现，相对于前面的手工集成 SSM 框架来说，它使用起来更加简单快捷。这就是 Spring Boot 框架的优势所在。

8.2　Maven 构建工具

在前面的开发中，我们一直使用 Gradle 作为构建工具。这一节，我们使用 Maven 作为构建工具。这两个构建工具比较起来，Maven 目前由于历史原因在业界使用得较多，而 Gradle 作为后起之秀，使用稍少，但从配置上看更为简洁。两者的共同点是可以使用相同的中央仓库。

8.2.1 Maven 简介

Maven 除了具有超强的程序构建能力，还提供了高级项目管理工具。我们所使用的 IDEA 平台就内嵌了 Maven。由于 Maven 的默认构建脚本有较高的可重用性，所以常常用两三行 Maven 构建脚本就可以构建简单的项目。由于 Maven 的面向项目的方法，许多 Apache Jakarta 项目发文时使用 Maven，而且公司项目采用 Maven 的比例在持续增长。

Maven 的项目对象模型(POM)，可以使用一小段描述信息来管理项目的构建、报告和文档。

8.2.2 Maven 的安装与配置

步骤 1．访问 Maven 官网。Maven 软件的官网下载链接为

```
http://maven.apache.org/download.cgi
```

步骤 2．下载并解压 Maven，结果如图 8-8 所示。

› 此电脑 › 文档 (E:) › 软件下载 › apache-maven-3.1.1-bin › apache-maven-3.1.1

名称	修改日期	类型	大小
bin	2017/3/16 11:59	文件夹	
boot	2017/3/16 11:59	文件夹	
conf	2018/3/5 9:51	文件夹	
lib	2017/3/16 11:59	文件夹	
LICENSE	2013/9/17 11:24	文件	15 KB
NOTICE	2013/9/17 11:24	文件	1 KB
README.txt	2013/9/17 11:19	文本文档	3 KB

图 8-8　Maven 解压的结果

步骤 3．设置 Maven 环境变量 maven_home 的值为解压的主路径。如图 8-9 所示，系统变量里添加了变量名 maven_home，值为 E:\软件下载\apache-maven-3.1.1-bin\apache-maven-3.1.1。

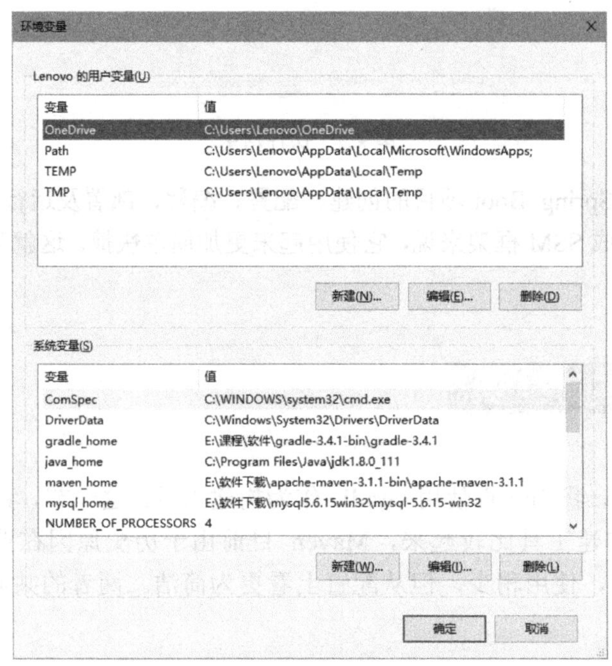

图 8-9　Maven 环境变量设置

设置 path 路径，如图 8-10 所示，添加 E:\软件下载\apache-maven-3.1.1-bin\apache-maven-3.1.1\bin 路径。

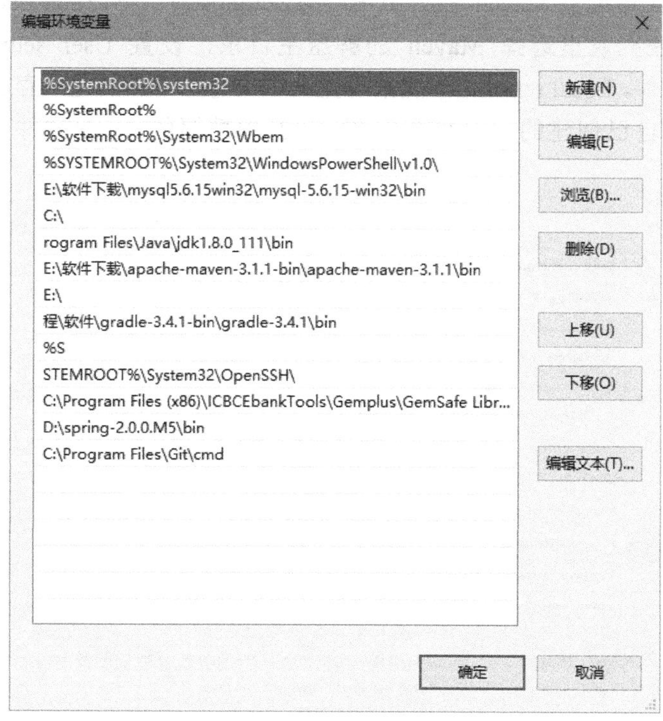

图 8-10　编辑环境变量 path

步骤 4．测试 Maven 的安装。如图 8-11 所示，在控制台下运行 mvn -version，显示 Maven 的安装路径和 JDK 版本等信息，表示安装成功。

```
C:\Users\Lenovo>mvn -version
Apache Maven 3.1.1 (0728685237757ffbf44136acec0402957f723d9a; 2013-09-17 23:22:22+0800)
Maven home: E:\软件下载\apache-maven-3.1.1-bin\apache-maven-3.1.1\bin\..
Java version: 1.8.0_111, vendor: Oracle Corporation
Java home: C:\Program Files\Java\jdk1.8.0_111\jre
Default locale: zh_CN, platform encoding: GBK
OS name: "windows 10", version: "10.0", arch: "amd64", family: "dos"
C:\Users\Lenovo>
```

图 8-11　Maven 安装成功

步骤 5．修改 Maven 的配置文件。Maven 的配置文件 settings.xml 位于 E:\软件下载\apache-maven-3.1.1-bin\apache- maven-3.1.1\conf 下，在该文件中添加如下代码段，可增加国内仓库镜像站点，加快下载速度。

```
<mirrors>
   <mirror>
     <id>alimaven</id>
     <name>aliyun maven</name>
     <url>http://maven.aliyun.com/nexus/content/groups/public/</url>
     <mirrorOf>central</mirrorOf>
```

```
    </mirror>
  </mirrors>
```

步骤 6. 在 IDEA 中配置 Maven。打开 IDEA 的 Settings 对话框，搜索 maven，在右边 Maven home directory 下拉列表里选择 Maven 的解压主目录，设置 User settings file 为主目录\conf\settings.xml 文件。Local repository 用来设置从中央仓库下载的包所存放的本地仓库地址，通常来说，会填写自己创建的一个目录，图 8-12 中填写的是 E:\软件下载\apache-maven-3.1.1-bin\reps。

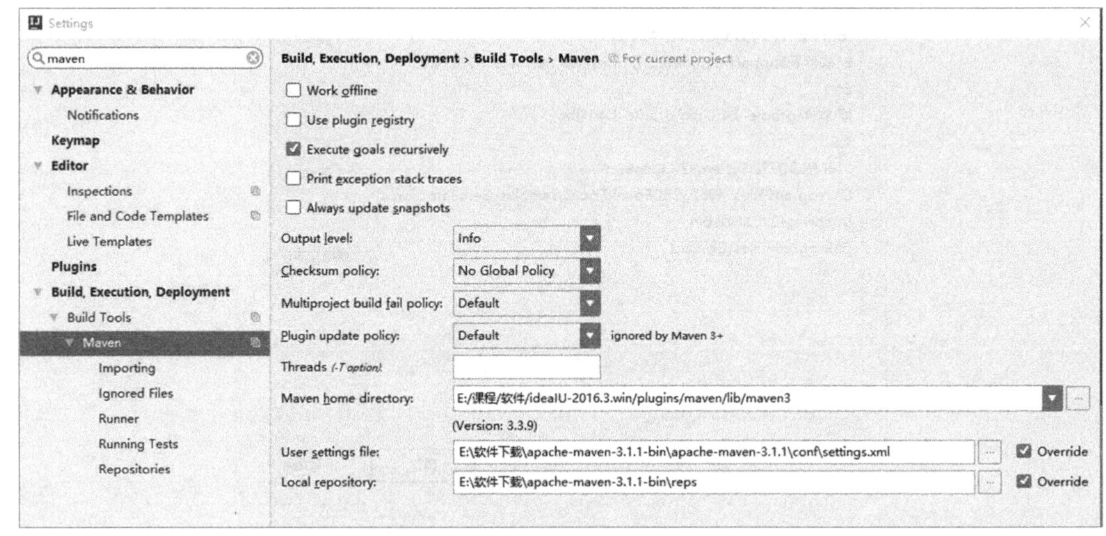

图 8-12　在 IDEA 中配置 Maven

提示：如果 IDEA 中 Maven 下载时中断报错。可以到仓库 E:\软件下载\apache-maven-3.1.1-bin\reps 删掉项目所依赖的包目录。然后，如图 8-13 所示，回到 IDEA 中，单击右侧边栏的 Maven Projects，单击左上角的刷新按钮，会重新下载项目。

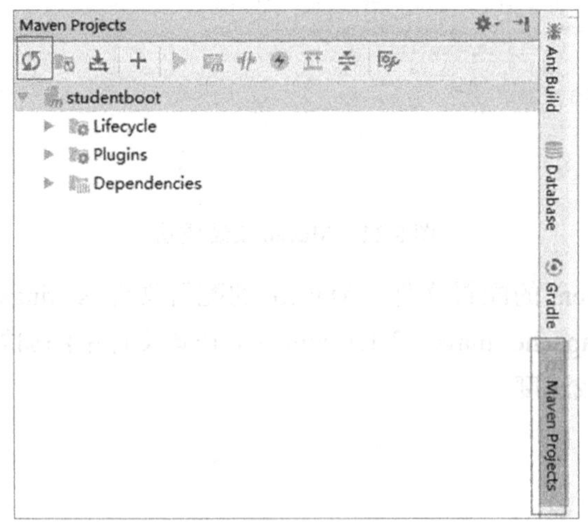

图 8-13　刷新 Maven 项目

8.2.3 pom.xml 文件

Maven 的 pom.xml 文件是项目对象模型管理文件，它与 Gradle 的 build.gradle 文件的作用类似，主要用于描述项目的配置文件，开发者规则，缺陷管理系统、组织和许可证 licenses，项目的 URL，项目的依赖性，以及其他所有的项目相关因素。大致的内容如下：

```
<?xml version="1.0" encoding="UTF-8"?>
<project xmlns="http://maven.apache.org/POM/4.0.0" xmlns:xsi="http://www.w3.org
/2001/XMLSchema-instance"
    xsi:schemaLocation="http://maven.apache.org/POM/4.0.0
https://maven.apache.org/xsd/maven-4.0.0.xsd">
    <modelVersion>4.0.0</modelVersion>
    <groupId>com.example</groupId>
    <artifactId>TestMaven</artifactId>
    <version>0.0.1-SNAPSHOT</version>
    <name>TestMaven</name>
    <description>Demo project for Spring Boot</description>
    <properties>
        <java.version>1.8</java.version>
    </properties>
    <dependencies>
        <dependency>
            <groupId>junit</groupId>
            <artifactId>junit</artifactId>
            <version>4.11</version>
            <scope>test</scope>
        </dependency>
        <!-- https://mvnrepository.com/artifact/javax.servlet/jstl -->
        <dependency>
            <groupId>javax.servlet</groupId>
            <artifactId>jstl</artifactId>
            <version>1.2</version>
            <scope>provided</scope>
        </dependency>
    </dependencies>
    <build>
        <plugins>
        </plugins>
    </build>
</project>
```

该文件里，依赖有个子标签<scope>，它的值跟 Gradle 类似，有 4 种有效范围，即 test、compile、provided、runtime，分别表示这些依赖包的有效范围是测试、编译、编译&测试、运行。

8.2.4 任务二：用 Maven 构建项目

在 8.1.2 节的任务一中开发 Spring Boot 框架时我们使用 Maven 构建了项目。除此之外，我们还可以跟 2.3 节使用 Gradle 构建项目一样，创建 Maven 的 Java Web 项目。步骤如下。

选择菜单栏中的 File→New→Module→Maven。如图 8-14 所示，在弹出的对话框的左侧列表框中选择 Maven，在 Module SDK 下拉列表中选择安装好的 Project SDK（1.8），勾选 Create from archetype（从原型创建）复选框，选择 org.apache.maven.archetypes:maven-archetype-webapp 原型，会自动创建 Java Web 项目的目录结构。

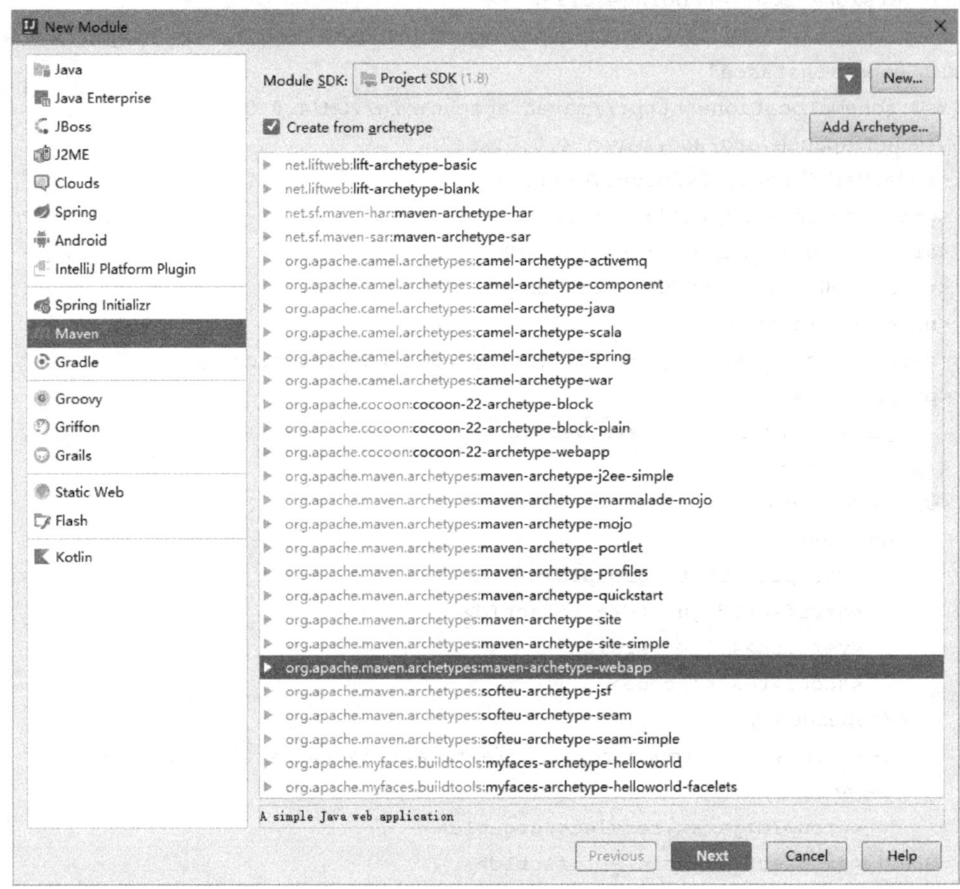

图 8-14　Maven 原型选择

单击 Next 按钮，弹出如图 8-15 所示的对话框，在 GroupId（项目组织名）输入框填写 person.xjl，ArtifactId（项目成品名）输入框填写 MavenTest，版本不变，单击 Next 按钮。

弹出如图 8-16 所示对话框，因为我们前面配置过 Maven，所以里面的信息都已经选好，使用用户自定义的 Maven 仓库。

单击 Next 按钮，弹出如图 8-17 所示对话框。在该对话框中指定项目名称和存放路径，然后单击 Finish 按钮，等待 Maven 工具自动下载原型并创建项目结构目录。

当 IDEA 界面最底部的状态条不再闪烁时，弹出如图 8-18 所示信息，则表明项目创建成功。

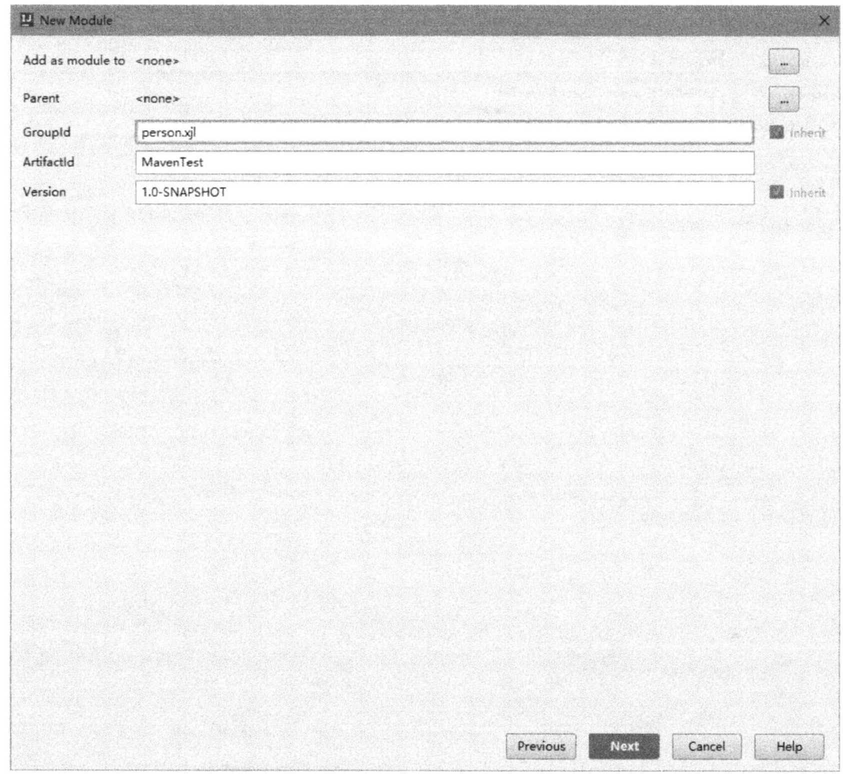

图 8-15　设置项目组织名和项目成品名

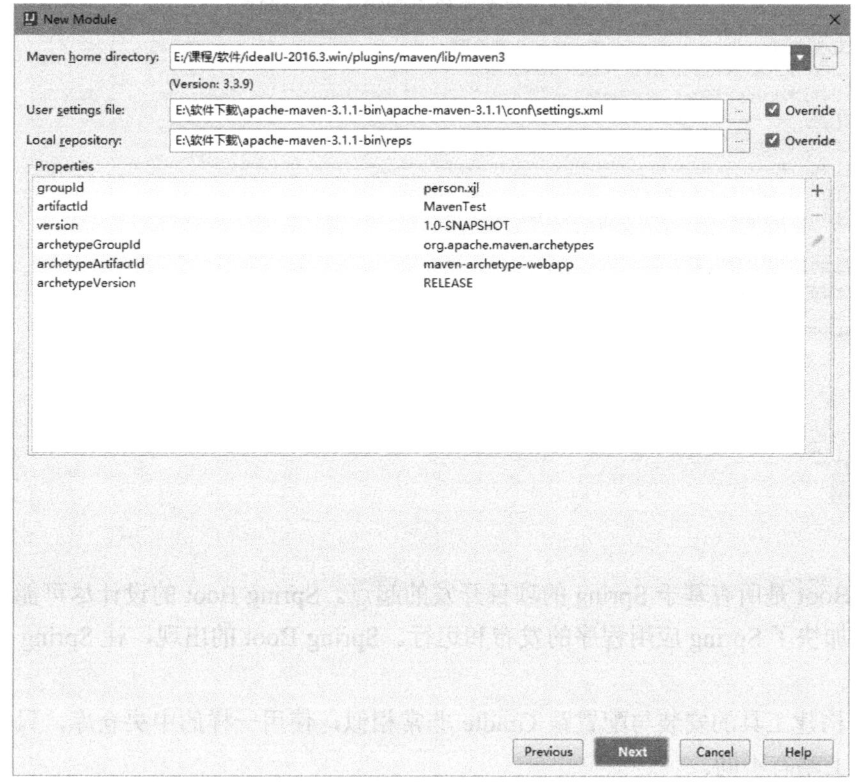

图 8-16　项目的 Maven 配置

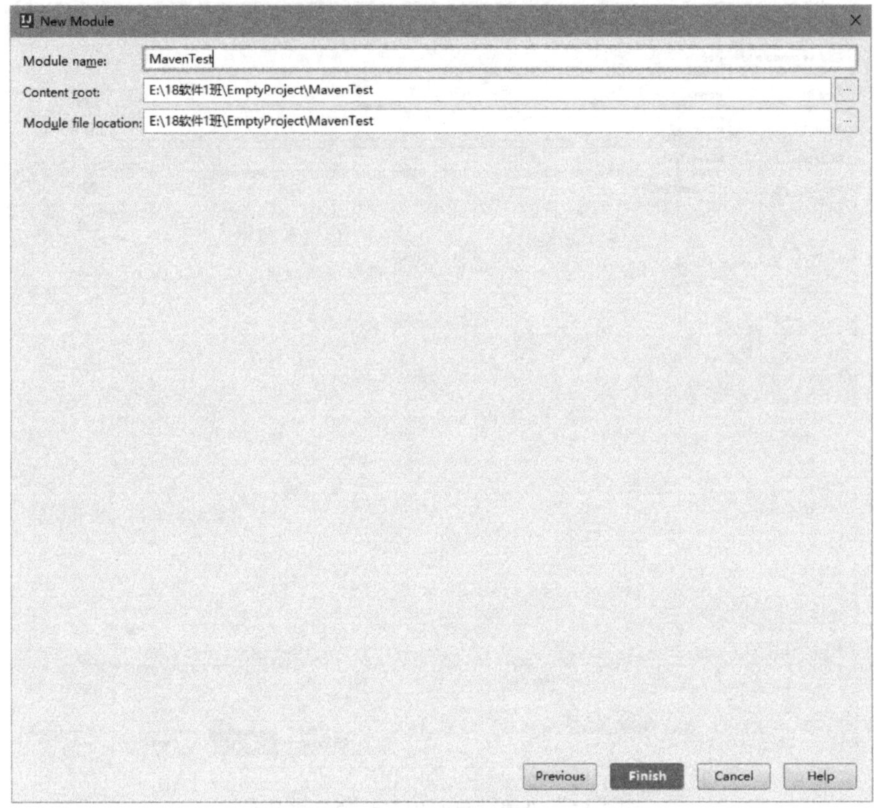

图 8-17　为项目命名和指定存放路径

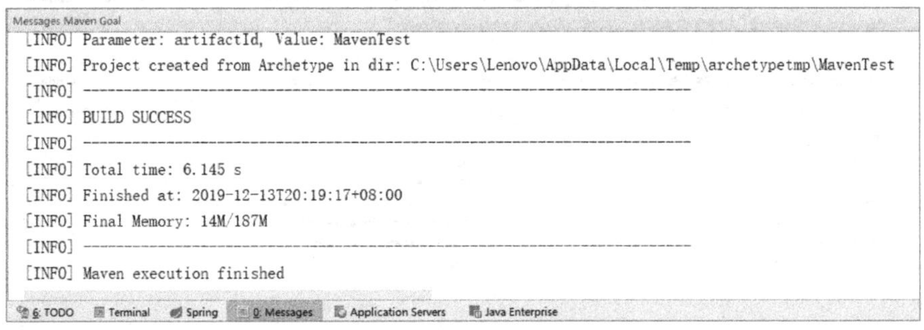

图 8-18　Maven 项目创建成功

小　结

Spring Boot 是所有基于 Spring 的项目开发的起点。Spring Boot 的设计尽可能地减少了配置文件，并加快了 Spring 应用程序的发布和运行。Spring Boot 的出现，让 Spring 开发进入快速开发时代。

Maven 构建工具的安装与配置跟 Gradle 非常相似，使用一样的中央仓库，只是在使用插件运行任务上略有不同。

习　题

一、填空题

1．从本质上来说，________就是一些库的集合，它能够被任意项目构建工具所使用。

2．Maven 的________，可以使用一小段描述信息来管理项目的构建、报告和文档。

二、简单题

1．Spring Boot 是如何管理依赖包的？

2．比较构建工具 Maven 与 Gradle 的异同点。

综合实训

实训 1．使用 Spring Boot 实现示例 5-2、示例 5-3、示例 5-4 对数据库的增、删、改的操作。

实训 2．使用 Spring Boot 实现示例 5-5 中登录功能。

附录 A

本书使用的数据库一共有 5 张表，具体内容如下：

（1）管理员表：admin。

（2）学生表：student。

（3）教师表：teacher。

（4）课程表：course。

（5）成绩表：score。

这 5 个数据库表的设计如表 A-1～表 A-5 所示。数据库表之间的关系如图 A-1 所示。

表 A-1　管理员表（admin）结构设计

字 段 名	字 段 类 型	是否允许为空	Default	主　键	Extras	备　注
id	int(11)unsigned	No	Auto_Incremnt	Yes	unique	管理员编号
username	varchar20)	Yes	Null			用户名
password	varchar20)	Yes	0			密码
name	varchar20)	Yes	Null			真实姓名

表 A-2　学生表（student）结构设计

字 段 名	字 段 类 型	是否允许为空	Default	主　键	Extras	备　注
id	int(11)	No		Yes	unique	学号
name	varchar20)	Yes	Null			姓名
password	varchar20)	Yes	0			密码
sex	int(11)	Yes	0			性别
clazz	varchar20)	Yes	Null			班级
birthday	varchar20)	Yes	Null			出生日期

表 A-3　教师表（teacher）结构设计

字 段 名	字 段 类 型	是否允许为空	Default	主　键	Extras	备　注
id	int(11)	No		Yes	unique	教工号
name	varchar20)	Yes	Null			姓名
password	varchar20)	Yes	0			密码
sex	int(11)	Yes	0			性别
course_id	int(11)	Yes	0			课程号
birthday	varchar20)	Yes	Null			出生日期
professional	varchar20)	Yes	Null			职称

表 A-4　课程表（course）结构设计

字 段 名	字 段 类 型	是否允许为空	Default	主　键	Extras	备　注
id	int(11)	No		Yes	unique	课程号
name	varchar20)	Yes	Null			课程名
teacher_Id	int(11)	Yes	0			教工号

表 A-5　成绩表（score）结构设计

字 段 名	字 段 类 型	是否允许为空	Default	主　键	Extras	备　注
id	int(11)	No		Yes	unique	
student_id	int(11)	Yes	0			学号
course_id	int(11)	Yes	0			课程号
score	double	Yes	0.0			成绩

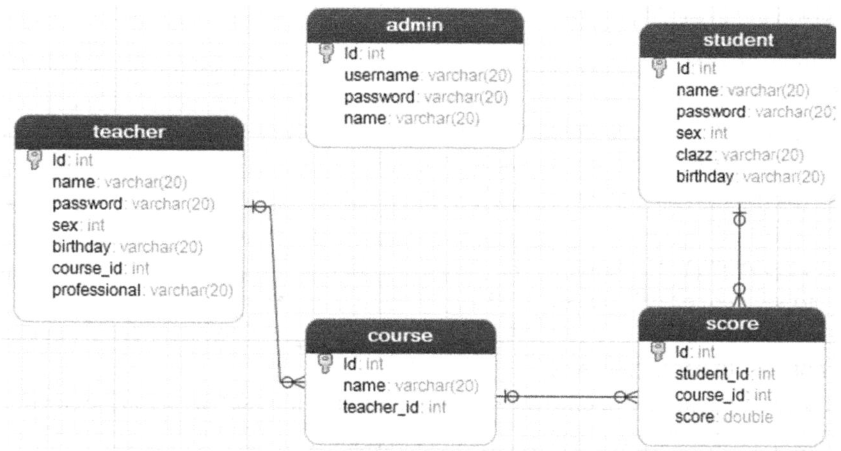

图 A-1　数据库表之间的关系

数据库创建的 DDL 脚本如下：

```
SET FOREIGN_KEY_CHECKS=0;
-- ----------------------------
-- Table structure for `admin`
-- ----------------------------
DROP TABLE IF EXISTS `admin`;
CREATE TABLE `admin` (
  `Id` int(11) NOT NULL auto_increment,
  `username` varchar(20) default NULL,
  `password` varchar(20) default NULL,
  `name` varchar(20) default NULL,
  PRIMARY KEY  (`Id`)
) ENGINE=InnoDB AUTO_INCREMENT=2 DEFAULT CHARSET=utf8;
-- ----------------------------
-- Records of admin
-- ----------------------------
INSERT INTO `admin` VALUES ('1', 'admin', 'admin', '超级管理员');
-- ----------------------------
```

```
-- Table structure for `course`
-- ----------------------------
DROP TABLE IF EXISTS `course`;
CREATE TABLE `course` (
  `Id` int(11) NOT NULL,
  `name` varchar(20) default NULL,
  `teacher_id` int(11) default NULL,
  PRIMARY KEY  (`Id`),
  KEY `teacher_course` (`teacher_id`),
  CONSTRAINT `teacher_course` FOREIGN KEY (`teacher_id`) REFERENCES `teacher` (`Id`)
ON DELETE CASCADE ON UPDATE CASCADE
) ENGINE=InnoDB DEFAULT CHARSET=utf8;
-- ----------------------------
-- Table structure for `score`
-- ----------------------------
DROP TABLE IF EXISTS `score`;
CREATE TABLE `score` (
  `Id` int(11) NOT NULL auto_increment,
  `student_id` int(11) default NULL,
  `course_id` int(11) default NULL,
  `score` double(6,1) default NULL,
  PRIMARY KEY  (`Id`),
  KEY `stu_score` (`student_id`),
  KEY `course_score` (`course_id`),
  CONSTRAINT `course_score` FOREIGN KEY (`course_id`) REFERENCES `course` (`Id`) ON
DELETE CASCADE ON UPDATE CASCADE,
  CONSTRAINT `stu_score` FOREIGN KEY (`student_id`) REFERENCES `student` (`Id`) ON
DELETE CASCADE ON UPDATE CASCADE
) ENGINE=InnoDB AUTO_INCREMENT=5 DEFAULT CHARSET=utf8;
-- ----------------------------
-- Table structure for `student`
-- ----------------------------
DROP TABLE IF EXISTS `student`;
CREATE TABLE `student` (
  `Id` int(11) NOT NULL,
  `name` varchar(20) default NULL,
  `password` varchar(20) default NULL,
  `sex` int(20) default NULL,
  `clazz` varchar(20) default NULL,
  `birthday` varchar(20) default NULL,
  PRIMARY KEY  (`Id`)
) ENGINE=InnoDB DEFAULT CHARSET=utf8;
-- ----------------------------
-- Table structure for `teacher`
-- ----------------------------
DROP TABLE IF EXISTS `teacher`;
CREATE TABLE `teacher` (
  `Id` int(11) NOT NULL,
```

```
    `name` varchar(20) default NULL,
    `password` varchar(20) default NULL,
    `sex` int(11) default NULL,
    `birthday` varchar(20) default NULL,
    `course_id` int(11) default NULL,
    `professional` varchar(20) default NULL,
    PRIMARY KEY  (`Id`),
    KEY `course_teacher` (`course_id`),
    CONSTRAINT `course_teacher` FOREIGN KEY (`course_id`) REFERENCES `course` (`Id`)
ON DELETE CASCADE ON UPDATE CASCADE
  ) ENGINE=InnoDB DEFAULT CHARSET=utf8;
```

参考文献

[1] 牛德雄，杨玉蓓．Java EE（SSH 框架）软件项目开发案例教程．2 版．北京：电子工业出版社，2018．

[2] 陈雄华，林开雄．Spring 3.0 就这么简单．北京：人民邮电出版社，2013．

[3] 温昱.软件架构设计．北京：电子工业出版社，2012．

[4] 库俊国．基于 J2EE 技术的 Web 应用体系研究及实践[M]．北京：人民邮电出版社，2014．

[5] 缪忠剑．基于 Spring 的集成化 Web 开发平台的研究与实现[M]．北京：机械工业出版社，2013．

[6] 王嘉．基于开源框架的在线学习平台的研究与应用[M]．西安：西安电子科技大学出版社，2013．

[7] 方志朋．深入理解 Spring Cloud 与微服务构建．北京：人民邮电出版社，2017．